The R.A.M.S. Library of Alchemy

Volume 31

The Metamorphosis of the Planets

by
Johannes de Monte-Snyder

R.A.M.S. Publishing Company

The Metamorphosis
of the
Planets

by
Johannes de Monte-Snyder

Produced by

Restorers of Alchemical Manuscripts Society

R.A.M.S. Publishing Company

R.A.M.S. Publishing Company
117 Rutherford Lane
Stuarts Draft VA 24477

The Metamorphosis of the Planets

First Edition 2015

ISBN-13 **978-1511733281**
ISBN-10 **1511733284**

Image Processing by Philip N. Wheeler

This book is sold for informational purposes only. Neither the publisher nor the editor shall be held accountable for the use or misuse of the information in this book.

Printed in the United States of America

Table of Contents

Dedicated to Hans W. Nintzel,

American Alchemist

and

Founder of the

Restorers of Alchemical Manuscripts Society

(R.A.M.S.)

Disclaimer

Liability: The publisher does not warrant or assume any legal liability or responsibility for the accuracy, completeness, or usefulness of any information, apparatus, product, or process disclosed. The publisher makes no representation as to the accuracy or completeness of the contents of this book and specifically disclaims any implied warranty of merchantability or fitness for a particular purpose. No warranty may be created or extended by written sales materials or sales representatives. You should obtain professional consultation where appropriate. The publisher shall not be liable for any loss of profit or other commercial or personal damages, including but not limited to special, incidental, consequential, or other damages.

Introduction

Philip N. Wheeler

Written in the 17th century, this extended allegory by Johannes de Monte-Snyder is indeed a puzzle. It perplexed Sir Isaac Newton for many years, as documented in his notebooks. As an allegory it links together the spiritual side of alchemy with actual physical laboratory processes. Snyder was an active Alchemist who practiced all over Europe.

Hans W. Nintzel selected this work for inclusion in the R.A.M.S. Library.

Adam McLean offers an extensive course in the private study of this work.

the METAMORPHOSIS *of the* PLANETS
John de Monte Snyders

The Metamorphosis of the Planets

A WONDERFUL TRANSMUTATION OF THE PLANETS AND
METALLIC FORMES INTO THEIR FIRST ESSENCE (WITH
AN ANNEXED PROCESS) BEING A DISCOVERY OF THE
THREE KEYS PERTINENT TO THE OBTAINING OF THE
THREE PRINCIPLES, LIKEWISE IN WHAT MANNER THE
MOST GENERAL UNIVERSAL IS TO BE OBTAINED IS IN
MANY PLACES OF THIS TREATISE DESCRIBED BY:

Johannes de Monte Snyder

The Preface.

Friendly Reader,

Since it hath pleased almighty God now in these times to shower down his grace as well upon the unjust as the just, and yet he hath resolved to spread abroad his mysterious and hidden Wisdom over the whole world, and to the end hath sent out his holy Angels to awaken as it were by sound of Trumpet the spirit of all true Philosophers, who hitherto have rested as it were bound up and fettered in a deep sleep, and have been concealed in the innermost recess of this world, with an express command to lay the same spirit upon others. So it is apt the time is now come, not that the deceased Philosophers should visibly arise in their bodies, but only that their spirit should be awakened, and of grace be layed upon others, whereby the goodness of the most high may be known in all the world; and wisdom may be made known to all (without distinction) who are willing to accept of her, without regard to their merits.

Now forasmuch as at this time the lot is fallen upon me, of all the most unworthy, I am constrained without delay to accomplish in sincerity and truth the will of the most High. To this End therefore in the name of my God, I as a seedsman am minded to sow the universal seed which the philosophick spirit hath imparted unto me, over all the world or stony, sandy, bushy, fruitful, and unfruitfull fields, so that neither the just nor the unjust can have any pretence to complain against God in this time of grace. And that I may now certify and make known before all the world my vocation that by the appointment of God the spirit of all Philosophers speaketh through me, and that I may demonstrate that through the spirit of God the Keys both of heaven and hell are committed unto me; it is very necessary that I bring to light the matter itself, together with the opening and shutting Key, whereby the mystery of this world, the womb of all natures is opened, all things are impregnated and promoted towards the birth.

But before I demonstrate my vocation by the characteristic testimony of all the Planets, I will first for a declaration of my already published treatise concerning the magical Elements, discover a most weighty point concerning the three Elemental irrational and mineral worlds, wherein I shall with most certain truth properly and significantly delineate the due matter, intreating the gentle reader not to be affrighted from the reading of the book by the small authority and slenderness of my person, but with all his faculties to take special notice of the figure and not of the thing figured. For I here place three irrational worlds wherein all depends, whereof one hath brought forth the other two. I set my left foot upon one birth of the great world, and place my right foot upon the other hermaphroditic world and its figure as a universal irrational and mineral microcosm; and I sit with my hinder parts on the mother of them both, which is the globe of the earth. And thus sitting I stretch out my arms towards midday and midnight, and touch with my hands both the great luminaries of the crystalline glittering heaven, and I swear by the most high Creator of all things that I have heretofore in my treatise written the truth, and will now in this description further bring to light that corner stone whereon so many thousands have stumbled and been scandalized, at which all true Philosophers have hinted:

I say true Philosophers in as much as the writings of many deceitful and false people have been accounted and esteemed true, or at least plain and clear, when nevertheless the greatest part know nothing at all, or at least have written but very little which might be advantageous to the seeking or erring lovers of Art.

And although I am not forbidden to point out these even by their very names, yet that I may avoid giving too great offence I shall forbear, and only busy myself in delineating unto thee not only the true matter through the three worlds, but also afterwards prefigure through the diminution of the forms, and permutation of the Planets, and likewise the whole consequence, as I in truth found it, whereof I shall not now boast. Wherefore turn thee unto the truth and lay aside the multiplicity of thy books. For the true philosophick books are depraved and counterfeited by many thousand venders of processes, glossers, and false Philosophers; and if one would take the pains to examine all with diligence, one should find no matter which hath not been forbidden or rejected by them. Wherefore if thou hast a mind to remain secure, then set thy mind on me, and behold me in my form and figure, credit my words which I have confirmed with this unheard of oath. For whereas with my arms stretched out I sat on the circle of the world, and with my hands or fingers pointed at the Sun and Moon, the true mysterious heavenly and earthly key is delivered to me and thee, and leave at will and pleasure to unlock the solary and lunary gates, and to translate the same from their office, honour and dignity, yea after a wonderful manner to eclipse them.

But whereas I with both my feet did touch the two irrational children of the great world, I thereby gave to understand that the nethermost was as the uppermost which I shewed with my fingers, and yet the uppermost was like the nethermost; and that the earth whereon with arms out I thus sat, was the mother and bearer of temporal salvation, who had subjected and made serviceable to me both her children, whereof the one was lunary, the other hermaphroditic; viz. a solary and Lunary birth. Both these are distinguished from each other by a small little stream laterally flowing out. And like as the Sun is to be discerned from the Moon, even so also are these two easily to be discerned one from the other, especially if it be considered that the Philosophers have appropriated to one birth the signs of water as a lunary character. For the Moon governs the Water and where ever the wise have adjoined a watery sign, the same is to be accounted Lunary. Concerning these three or rather two worlds I have taken upon me first of all to treat and afterwards (to manifest my vocation) of the permutation of the planetary forms, of their corruption, generation and melioration of the whole Essence with the greatest brevity that may be.

He that hath ears to hear let him hear, and he that hath eyes to see let him see and search, and so he will here in this place find that very thing which hath been elsewhere sought by many thousands and yet not found.

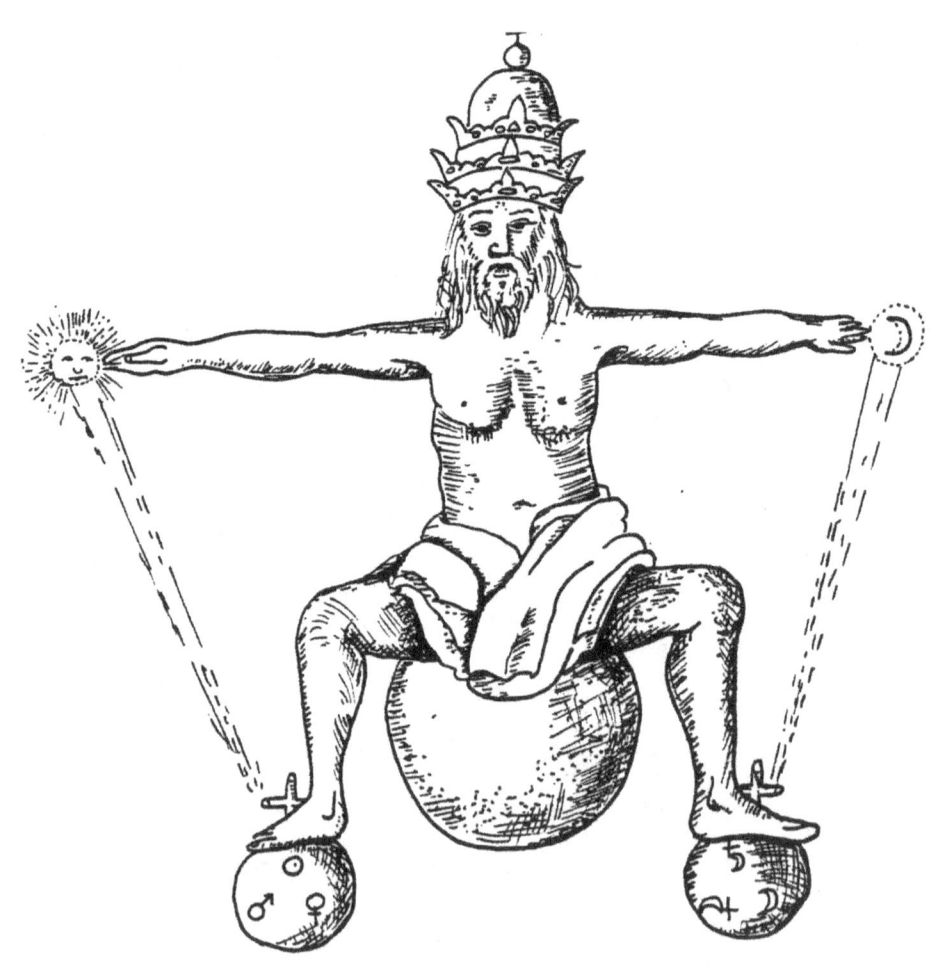

THE METAMORPHOSIS OF THE PLANETS

CHAPTER I

Concerning the Three Worlds

Like as all elementated things are distributed into
three kingdoms, even so also are there three
irrational worlds to be considered philosophically,
for that which is above is as that which is below,
which proceeds hence in that the Almighty in the beginning created the transparent heaven in a round
perfect circular form, and yet the same heaven in
like manner was framed of such a nature as to
produce its like, and yet according to the command
of God with most perfection: From whence there are
now arised and in the heaven spring forth two
luminaries, which daily and mightily do present
before the eyes the figure and character of the
circular crystalline heaven, and do as it were
testify and make known their original.

And like as the Almighty creator by the creation of
the heaven hath delineated himself in the celestial
world, and that in like manner by two other

luminaries hath produced the image of its creator; so also hath God depainted himself in the inferiour Elemental world; which rude world even as the upper world originally sprang forth in a circular form through the word Fiat. And that these might in all things conform themselves unto the superiour worlds, the rude Elemental world hath copied or pictured out itself into two little worlds, and produced its like.

The product is masculine and feminine, even as in the celestial world both the great luminaries as such proceeded from the command of the Almighty. The Sun is masculine and diurnal, the Moon is feminine and nocturnal. Now as all the celestial bodies participate and receive light from these two luminaries, just so do all Elementated things in this inferiour world partake of the masculine and feminine seed, specially the metals, which in their mines were formed by the masculine and feminine product, which in my little treatise I name the two generated irrational Microcosms, in as much as the same are marked with the character and figure of the great world their Mother, which by the command of God bore them. The earth hath not only a living spirit but also a living soul which is her seed wherewith she is impregnated and stirred up to produce her like. Now the more exactly any offspring agrees with its mother in form, figure, virtue and property, the more noble is the same to be reputed. For Parents by an excess of singular love are wont to mould and fashion their own children one of them before the other, and as it were to impress and figure themselves in them, so that the Parents, especially the Mother amongst some children cannot be distinguished. Whatsoever is born in an ordinary way proceedeth from Male and Female, especially in animals.

From one God the Archetype sprung forth in the heavens his luminaries, which do infuse into the rest of the stars their light and brightness and do also promote the same unto their predestination. Heaven then is an exact image of God, and the two luminaries were produced and created after the similitude of the image of the divine Majesty. And that I may evidence that the nethermost is even as the uppermost, know that even God Almighty after the same manner hath represented himself first of all here in this elemental world in as much as he created the world in a round form, and as it were a seal of his perfection and eternity, which would again by virtue of the command, Increase and Multiply, hath brought forth two little worlds which bear the symbol of the creator. For no creature considering it in a higher and lower degree is without a character of its Author, and in the same manner it agrees with that form from which it was derived. Now the one is a Lunary female and nocturnal governess, but the other is masculine solary and a ruler of the say. Both these were created, not after the similitude of God but after the image, viz, the similitude of the world. So likewise there sprang from Adam two sexes of whom all mankind is born. Yea God himself who from eternity was without Mother was born into the World without Father. And that he might thereby the better corroborate this ordinary way of generation, he was from eternity from the Father but in time was born man and became flesh from a Mother, so that two Natures do individually subsist in the second person of the Diety. And after the same manner that two natures do exist in the second person, there are likewise two natures to be observed in one only birth of the great world, so that in one only Hermaphrodite may be found perfectly, by way of concentration, all Mineral, Vegetable and Animal powers, virtues and properties. This birth hath of itself a true symbol with superiors, for it hath a likeness with the Sun and Moon, for as much as even gold and silver are produced from the same. It hath also plentifully received universal influence from the great intelligence which shall be here specially considered. Now it seems to me that it hath been clear enough spoken concerning these three worlds, as also concerning this, yet the superior is like that which is inferior, and contrary-wise: Wherefore I say unto thee again that God hath delineated himself in the earth, yet is the world, and the earth hath again figured itself through its two births or worlds, namely through two minerals. He who knows not these two minerals is also ignorant of the great world and its Mother. The parents live in their children, and by the parents the children are known and contrariwise. But how this knowledge is to be obtained the courteous reader will find in the sequel, which is to be understood after a philosophick manner, and not otherwise. For even as the philosophick Mother is, such also is her birth; and as the tree, such is its fruit. For everything is naturally fitted to produce its like, whence a good tree brings forth good fruit. And corn as in the seed of a tree the intire nature, and property of the tree is concentrated, so principally in one thing namely in our birth of the world whole nature is concentrated. This is the due and next matter and etc. and is called the concentrated metal and mineral of which in the following chapters I shall treat further. But one thing I shall here put thee in mind of, which is specially to be noted, concerning the double nature, namely of the one hermaphroditick birth of the great world:

That like as in the seed of all things the nature
and property of all things is driven into a narrow
compass, even so the property of all the world is
visibly impressed by a peculiar and proper character
upon this new born hermaphroditick child, by virtue
whereof the child bears the likeness and figure of
its mother inwardly and outwardly, and is named the
next matter. He who understands not this my meaning
is as yet to be esteemed no Philosopher:

For whoever is ignorant of the root of minerals
knows not nature's principles. By the root
understand the seed of all metals and etc. which
without root can bring forth no fruit. But how such
a seed of two-fold nature or birth is to be
discerned, shall be clearly taught in that which
follows. Wherefore look for it.

CHAPTER II

How to know the world from the world: That is, How
the Hermaphroditick little irrational mineral world
is to be distinguished from the Lunary feminine
birth. And farther how the nethermost is like the
uppermost.

The cold Earth first delineated itself in a cold
waterish mineral, which by modern fools is held to
be a metal. But I say here that the same cannot be
metallick but mineral, and this mineral according to
the disposition of the mine or quality of the matrix
is Lunary. For this birth wherein the great world
hath copied forth itself, is even as Saturn a
producer of silver. For the half-moon above and
beneath attesteth her proper predestination. This
material lunary little world containing in herself
running mercury, which as a little stream by the
side of the diameter flows therefrom. The small
streams and the water is characterized of the
Philosophers like little streams by a diametrical
line. The sign of water denotes mercury which is the
primordial water of metals, that hath taken in the
whole body. And although the same is outwardly
adorned with a beautiful red yet it is indowed with
very little solary sulphur. Wherefore the matter is
Lunary, flegmatick and likened to a Queen, who

notwithstanding her monthly red flux, must continue in her sex until the appointed time. This is the first born daughter of the world who is concealed in the hollowness and clefts of rocks, and by peculiar accident was divided from the breasts of her mother. The great world, notwithstanding she had brought forth this daughter, still remained great with child, and brought to light a solary masculine birth, which is to be reckoned as an Hermaphroditick and double nature, for as much as from and out of it both sexes, viz, the Moon and the Sun may be born. This birth is likewise that first requisite entity of Moon and Sun, in which the restauration of all Minerals, Vegetables and Animals is perfectly comprehended. He who now understands to distinguish both the children or worlds, and can use each of them according to its predestination, the same not only knows the metallick water, but to him also is manifest the true Virgins Milk and Sulphur of Sulphur. Neither can the fixt Salt of the Earth escape him. And he will so mix the sulphur and Salt with the living metallick water, that they can never be separated from each other and etc.

Each thing loveth and by the ordinance of God produceth a thing like itself. This I first shewed in the Archetype, who after he had made the earth in a perfect circular figure, thereby to portray his perfection and infinity, did therewith give command to the world to produce two other little worlds as its like, as hath been more largely declared above. The heaven after that itself had by the will of God been made like him, brought forth also two luminaries in its circle, whereof the one is Lunary, the other Solary. And even as God Almighty is in himself a light, so hath he manifested unto us this glance by the above mentioned Luminaries, in which he portrayed himself.

CHAPTER III

Of the double and universal nature of the Hermaphroditick little World.

This my doctrine will at first to many seem exceedingly wonderful, but if they give good heed to my writing they will not think it so very strange, although this in respect of the Lunary birth appear repugnant to my first treatise concerning the Magical Elements. For there I mentioned that the earthly Saturn or Lead contained and produced the seed of Silver and Quick-silver, whence in the Lead Mines in the very deepest of them much unripe Silver is found and etc. This is true and is not now gainsayed by me, but serveth here for an Elucidation of the former. That the common decrepid Saturn as a bastard of the true Saturn participated such a nature from the feminine Lunar Child of the world, and if so be the defective Saturns salt were thereto disposed, the Lead Mines which hold Silver might produce greater advantage. How it may well be concluded whence it proceeds that those who place their hopes upon the mercury of this Saturn, are deceived in their opinion. The mercury of Lead is not of itself sufficiently disposed to produce Lune or Sol, but either more or less according as it partaketh much or little of the true child of the world, Virgin Mercury, Sulphur, and fixed Salt in its mines. The common Mercury which springeth out of its own minera is cleaner and much more noble than the rude saturnal Mercury: For with this by the help of Venus and the true hermaphroditick spirit may for a better process be ordered then many conceited proud and haughty scribblers and process-venders will easily dream. There are divers processes and particulars in which amongst other ingredients the true matter of the work is also comprehended: But forasmuch as some other repugnant particulars are used therewith, the true matter cannot so as behoveth perform its office. But he that is indowed with a true Philosophick understanding, esteemeth no process but reforms the particulars as he pleaseth, so that, the same introduce (produce) much or little according as to him seemeth good. He who contemplateth the true matter or first Ens of Gold may find the same expressly described and portrayed throughout all the Planets in my treatise concerning the magical Elements: But he knows it from the Mother, of which the same was born. But this matter is such that the Philosophers in its proper place speak least of it. He that hath wrought many Processes or else read them throughly over let him only give heed to that matter which he hath least of found in his processes; whereof I have spoken in every leaf. For this say I (by my Faith) is the true

matter and the duplicate nature as to the spirit,
but according to the soul it is Solary. The
Philosophers have compared this matter to a three
headed Dragon for two reasons: First because by the
mediation of this matter all metallick bodies may,
yea and must be brought to their first matter,
namely into Mercury, Sulphur and Salt. Secondly
because in the preparation and decoction of the
Stone three chief or principal colours, viz, black,
white and red, are found. With these colours I say
they have adorned the three heads of the crawling
dragon. It is the universal matter and to be found
all the world over. For the world without ceasing
produceth such a primum Ens or double nature,
whereby the Philosophers have obtained health and
riches.

CHAPTER   IV

How the double nature or young hermaphroditick world
falls in love with the Universal property. And how
the same harmoniseth with Venus, and etc. and is
forsaken by her.

The Hermaphroditick new world is so clear and
express depainted before the eyes of all Lovers of
the Chymical Art, that it cannot be clearer done. O
thou honourable, astral, earthly, salt, moist, dry,
light, heavy minera, and chosen Electrum: How hast
thou sinned against the noble Venus, that she so
long has withdrawn her assistance from thou. Thou
art in love, and goest day and night at the
Alabaster and Ivory feet of the Cyprian Queen, Thou
endeavoureth with thy fiery eyes to inpress upon
Venus by way of fascination the inwardly burning
love and desire of thy heart and to make her
according to thy will of like form with thy self.
But noble Venus fleeth and despiseth thee in regard
of thy abominable shape, and uncouth opposite
figure, Thou manifest thy crass and discoverest thy
fiery distemper. But Venus does the contrary, she
bareth her head, and covereth her shame and body
with her green garment. O thou most unfortunate
general suitor! How hath love handled thee: How is
thy form faded? Thou art invested with inremediless

mortal bitter and ingrateful dolour. Thou art loaden with the greatest crass that can in this world be found. Thou woundest thy breast, and longest to refresh the thirsty with thy blood, as a Pelican her young ones. It is not enough for a Monarch or king to be a little king or petty Prince. But thou strippest thy self of thy royal ornaments, and thy purple diadem thou bestowest willingly, and therewith cloathest the naked setting them in thy kingdom. All have need of thee, and yet they hate thee, and return evil for good. But the king of all Planets which was born out of thy body he loves thee. Also his purified Earth, viz. Salt, strengthens thy soul and doubled spirit. The Lepers make hast after thee and desire to drink out of thy Beker. Thou purgest and healest their leprosy. Thou cureth diseases fundamentally and implantest permanent health. Thou exilest poverty and bringst riches and all hidden treasury to light. Thou art he that art come and shall come. Thou art the same through whom all are come and through whom also they must be brought backwards and again forward into their highest honour and dignity. In sum thou art the true one and no other; wherefore it is fit that thine acknowledge thee.

CHAPTER V

How all the Planets offer their services to the Monarch of this World, and after the common guise of the world pretend kindred in hope to obtain great favour. And how Mercury and Jupiter fell at strife about it.

This was noised throughout the whole world. Behold now each one will be nearest of kin. Each offers his hands. Mars and Venus come with their Peliga (pledges?) and will convince thee that they are of thy stock and family: For they bear part of thy arms and eithers Escutcheon is inwardly adorned with thy golden colour. Mercury stept in barefooted without the ornament of his head, bowed his head to the earth, humbling himself above all measure in hope to be exalted. He was accepted: For the most powerful who bears in his hands the law both spiritual and temporal, was formerly carried upon the wings of Mercury into the kingdom into the Throne, which is granted to none but him to possess. For essentialized Mercury as a wind carried him in his belly. By the wind is always the Air and by the Air Mercury understood.

The gray-bearded Jupiter understood from a Comet and signet-star, that the double nature as a Monarch of this world governed his kingdom in peace by the assistance of Mercury, and that from all parts of the world envoys were arrived to congratulate and complement the most mighty and invincible. Wherefore the good Jupiter mounted upon the wings of his nimble Eagle, and hastened to the Palace and having obtained audience steps in, makes his due reverence with his scepter, bows his knees, kisses the foot of the Monarch, presents his Eagle to his service, intreats like as all the rest also for an inheritance, and an eternal kingdom to be communicated unto him, in consideration that the old Dragon his royal Majestys Father, together with his own Father, the powerful Eagle, namely with Jupiter elect, by the counsel and good advice of the assembled estates and Burgesses of the Philosophic kingdom, had established an inviolable friendship. And therefore he desired, since at that time there was such friendship between the old Dragon and the powerful Eagle, which is also wont to be named Bismutum, that for the present also Mercury should no longer withhold his favour from him, especially since as he thought he was most serviceable to his Majesty, and arose from his blood. To manifest this he wounded his heart, and showed his bloody Solary Sulphur which he inherited from the old Dragon. This being done he requested to sit at the right hand of the Monarch. At that Mercury suddenly started up stepping into his circle, said: Dear Brother thou talkest indiscreetly and understandest not what thou desirest, and the foundation upon which thou groundest thy self is to thy own prejudice and my advantage. The Philosophers who erected this Monarchy by the assistance of the Eagle who was exalted by the Venereal property, do not then understand thy Eagle when perchance they speak of an Eagle, but they speak of a Mercury which is born out of a nobler nature before he became an Eagle. I am said he the true Eagle having feathers on my hands, head and feet, I am light and above all measure heavy, and am fraught moreover with all the qualities which the Philosophers have attributed to Mercury. I am like to the most high in shape. For when I suffered myself to be exalted, I promoted this Monarchy; and when I humbled myself, I was through the exaltation of my cross raised to his honour, and dignity. Whereupon dear Brother for this time be satisfied.

CHAPTER VI

How Jupiter and the changed Mercury strove one with another in presence of the Monarch: And how Mercury convinced Jupiter.

Jupiter could no longer refrain, but gnashing his teeth was ready to fall on with thunder and lightning. But the Regent of this world commanded peace and quietness, licensing Mercury fully to prosecute his plea. Whereupon Mercury spake. Gentle Jupiter be not wrath with me thy Brother, but rather call to mind that the Philosophers say, that in case one only Drachm of a Malleable body be to be found in our matter, that the same is unfit and not to be accounted the true matter. But now I remember, and it stands still fresh in my memory, that thou at the first entry and audience obtained, didst so crook and bend thy knees and whole body, that at that time I could not perceive any unmalleableness in thee. How wilt thou now here expidite thyself? Thou art porous and light, and little bettered of my spirit: This the Philosophers hint by the Eagle on which thou sittest and art carried up and down on. Thou and thy Eagle are two but I and my Eagles wings are but one. Then could Jupiter refrain no longer, stammered horribly, and insisted upon his possession, saying that at all times and even still he was named the most high, and that for the indication of his sublimity the Eagle was appropriated to him. For even as the Eagle amongst all the winged crew adventureth to fly highest, and approach with his young ones nearest the Sun, and therefore is named the king of all birds; so he was always to be esteemed the uppermost amongst his six breathren, in as much as he had always possessed the highest place in the Assemblies of the Gods and Planets, and administrated Justice to all supreme in the Councel. Mercury being no longer able to forbear laughing, stept out of his circle up to his Throne, and on him was set an imperial Crown. After which in his own Hungarian speech, he said: Gentle Brother, thou now dost even as all pervertness of the law are wont. Thou alledgest well, but pervertest the meaning of the law. Try thee now examine thine own nature. Thou art aery and light, but I am as the Dragon in ponderosity. We love one another in a most peculiar manner, having both of us metallick bodies. We were originally born of the same mother, namely the great world. I am the earthly black Eagle, and was heretofore washed with the corrosive of Neptune and by the Venereal property exalted into a most beautiful, crystalline, weighty essence. At that time I helped to exalt the Dragon to this honourable throne, where all his breathren worshipped him. Herewith Mercury made an end, and returned thanks to the supple and malleable Jupiter, that contrary to his own intention he was fain to punctually to reccon up his pedigree qualities.

In the meantime there came crowding in with his one foot the old decrepid and impatent Saturn, halting in great haste towards the circle; but in getting up he fell upside down, whereupon there was raised a mighty laughter amongst all the Planets. And Venus in particular (after the manner of young women) laughed and shouted at the old man's misfortune. For he scrambled with his hands, and with his stump-foot could not so much as raise himself up. In such manner had he overlaboured himself in mercuries circle, yet he could hardly fetch breath. Which notwithstanding he would needs straight produce his plea, and thus sitting upon the Earth with a most hoarse voice began to speak. He blamed all the Planets that they without his presence had congratulated, and by congratulating had as it were established the Monarch. He alledged many prejudices on purpose to retard and fully to annul the act. For he was ready to maintain that he as a prover of them all, must also approve whatsoever was concluded in the assemblies of the Gods. But as he was thus speaking the most inraged Jupiter interrupted him and spake with a loud clap of thunder, so yet Saturn was fain to be dumb being already hoarse and weary, and Jupiter spake to the Monarch and said, that it was enough for him that he had been so long in such quiet possession, and that he desired to be maintained therein; saying that Mercury himself according to his own confession by reason of his earthly coldness and pre-alledged ponderosity, ought worthily to be accounted for the undermost of all and not for the most high. But I said he in his mother English speech am spiritual, angelical and altogether divine, and of my own strength able (together with my young ones) to climb up to the circle of the Sun. I am Solary and fiery, whereupon it was granted me to bear about thundering and lightning forever. I ought saith he and will have my seat and even with the King. For I can guard the King, and therefore am called Jupiter elect, because I as a regent of thunder and lightning am able to stand by and defend the most high, and therefore it stands written:

Fire in the presence of the most high. How cannst thou now thou Vagabond fawning dissembler appropriate this worthiness to thyself.

Then answered Mercury like a nimble spirit in his Hungarian mother tongue, and said; Gentle Brother, thou with thy Thunder and lightning hast spoken exceeding well, but much beside the mark. Thou shalt understand that although I am cold, earthy and heavy, that I am nevertheless spiritual and celestial and an ever burning living fire, wherefore I am called living Mercury because I do as it were awaken move and enliven other things which are dead. In respect of these my laudable qualities I am used as a most upright Arbitrator between the highest and the lowest, between the heavenly and earthly: I take part with the uppermost, and give assistance to the nethermost and tie them both together with a most indissoluable bond of love. I as a double nature am familiarly acquainted with both of them, and am no fawning dissembler but a peace-maker. Here Mercury raised his voice and said: Gentle Jupiter, thou art solary just as the hellish Lucifer is divine, and even as fire has its self towards water, just so dost thou accord with the Sun. And that thou mayst know the reason of thy thundering, take notice, that the fire, which is in the second quality is Air, strives against thy first moist quality and flyeth. Whereupon the inclosed fire crackles if thou with thy moist proper first quality approachest, to the other fiery quality which is called a convenient quality. And if it be fire viz. Thunder and lightning were in thee, thou wouldest be forced as a lord of the watery triplicity to die before thy time, for two contraries without a mean cannot subsist in one subject. Material fire calcines and burns up thy residence, so that it can be enlivened again by no mean but by me the reformed Mercury. Wherefore thou oughtest not to appropriate to thyself the philosophic speeches, for they only concern me. I am the Eagle, I am the Philosophick Air, and am also the true Sal-Armoniack. I by my wings have brought the King of the Earth up to his throne, and through the help of the true fixt metallick salt have established and ordained an everlasting Kingdom, which hence forward shall be subdued by no Earthly enemies. I have declared my predestination, and on that behalf Mars, Venus, Luna, and Sol must bear me witness; and therefore I desire that the most high Emperour will decide this controversy.

CHAPTER VII

How Jupiter during a cessation of speech was, as it were, commanded to keep company with his Brother Saturn. And how Saturn by this fellowship and conjunction, made his musters, and obtains the regiment.

Now as soon as ever Mercury, as was above related, gave over-speaking, he be-took himself again to his station, and there was stillness in heaven and on Earth, and behold a spirit arose out of the Earth which breathed a certain heavy thick cloud toward the royal golden throne, and a certain darkness incompassed the same, and he that sat upon the throne spake out of the darkness. His words were weighty, fiery beams, which looked like a bloody offering, which was layed upon the Altar by Mars and Venus, and was enkindled by Vulcan. These words might very well be marked by the Oriental Uegarian and Swedish language. For the spirit of the double Uegarian nature was to be reckoned without the soul in regard of so mighty and heavy a spirit: Wherefore it demanded the soul of Mars and Venus as a convenient holocaust. But the result, which went to the heart of Jupiter, was this. Who is like the most high? Thou hast purposed to exalt thyself with craft and iniquity, wherefore thou shalt join thy self to his right halting side to afford him all service at his pleasure, and shalt also so long shut up his concave with thy concave till he likewise shall have procured audience to his plea. Hereupon Jupiter swooned and old Saturn stood in and upon his own circle without stilts or feet, having his arms stretched out like one that had been crucified, and then with his fiery eyes he gazed upon Venus, and then again upon Luna, but lastly he beheld himself, was astonished and saw not himself, for his shape while he slept was changed by the conjunction of Jupiter. Wherefore he said, behold I am become like unto the most High, as ye all will perceive and in me is accomplished the truth of all the wise, who have attributed the true work to me Saturn, where they said; The Coagulation of Mercury is found in Saturn, wherefore O Gentle Mercury, thou injurest thyself, for if thou hadst yielded me such assistance as thou affordedst him who at present sits upon the throne and is still adorned of all the Planets, I would truly have much better and more richly have rewarded thee than thy Monarch hath done. And behold I swear to thee by my figure and new acquired form that I will never suffer things to rest thus but will at once open my mouth and burn up thy Wings, thine I say who by the help of the Universal property and the evaporated Neptune art become an Eagle, and will at length wholly swallow thee down, and in due time as the Whale did Jonas, bring thee to light again, then shalt thou be transmuted into a white Swan. Thou hast called forth, the Dragon out of his Den, and hast exalted him above all the stars, but me thy brother Saturn, who am the uppermost of all the Planets, thou hast forsaken, yea, oppressed. I am the true Saturn, the ALPHA & OMEGA, I am a trier and justifier of all the Planets, I suffer nothing unclean with the Queen, neither with the King. I am the excellent ornament of all royal crowns, and the golden apple for which the war between the Greeks and Trojans arose. I will, said he, farther darken the Sun and Moon, pluck down the stars from their places and station, call the winds by their names, open the earth, and bring forth the dead, and inspire a living breath through my mouth into the true Philosophers deceased, and then suddenly learn of them of what Saturn they writ so many books. I am most righteously jealous of mine honour, and grant mine honour to no stranger. The time is already accomplished and the days are hard by that I shall be clarified amongst all the people of this world. I will command the Earth, and she through this spirit of her first feminine lunary birth shall carry me on high, and convert me into a wholesome fountain, in which Mars shall wash his bloody coat, and Venus shall loose her nightgown. With these I will cloak

myself, and shew myself upon the rainbow, from
whence I will call unto me the soul of the earth-
that the soul and spirit may come together. The
Earth I will burn up, and thence elixiviate a
superfixt body. With this body I will espouse
myself, and with it become a briny Sea, and thereby
the Sea shall become dry land, and the land a
transparent heavy fluxible stone and etc. When
Saturn had said this he turned himself towards
midnight, and there proceeded a heavy vapour out of
his ears, and out of his nostrils and mouth a most
black and stinking smoke. Whereupon there was great
darkness upon the world, and there was no more life
to be perceived in the world until he opened his
eyes from whence proceeded horrible flames of fire
mixt with strange and wonderful colours. Whereupon
the whole world together with all its treasures were
manifested, and I heard a confused noise of many
voices which I could not distinguish, and me thought
that Vulcan was ready to unite the innermost part of
old Saturn with the Soul of Mars and Venus together
with the superfixt body from the Sun. Whilst I thus
listened and beheld, the darkness vanished and I saw
by Moonshine from far Saturn sitting with a long
white beard, blue garments and a red coat
underneath. The longer I looked upon him, the more
strange and admirable he appeared to me: For he
never aboad in the same shape, until at last the Sun

45

arose under his feet, and the Moon rested just at
the top of Saturn's bald head, and Mars and Venus
accompanied him on both sides, Jupiter and Mercury
were retrograde, stood in empty places; for the
final sentence was not yet passed. Mean time Saturn
who before sat on the Rain-bow, but now was by
Vulcan united with his soul and salt, was cloathed
with the Sun; so that all the Planets standing by
were dazzled, until at last by a peculiar mean they
were rendered capable of partaking his lustre.

CHAPTER VIII

How the Philosophers awake out of their sleep and explain themselves, and how the limber Jupiter is condemned.

After that Saturn, but not without the help of Jupiter as is above recounted, by a singular conjunction was turned into the Saturn of Philosophers, and afterwards into the highest of this world and the salvation of all the Planets, to him thus sitting upon his Throne was transmitted the sword and scepter: Whereupon, as in the last instance, Mercury and Jupiter were cited. Both these bestowed themselves lustily, each resolved to drink the wine of sentence with his witnesses. For their actions were presented, and the baskets as is usual layed by in readiness, each one being between hope and fear. Meanwhile there came a great multitude of choice and reverend persons who gave a lustre from far. And as soon as I well saw them I saw that it was a company of Philosophers who came very handsomely and orderly forward. In the first rank was Hermes & Geber. Hermes had a Phoenix in his arms whereon stood written: That which is above is like that which is beneath, and on the contrary; Geber bore a Pelican adorned with this inscription in golden letters: In the Sun and the Salt of Nature are all things. Morien with his two serpents each of which took hold of the others tail and so formed a circle, stept very thoughtfully forwards and spake with a clear piercing voice: Let the hidden be made manifest that we also may be manifested. Next to him came one characterized with a Basilisk, and me thought it was Roger Bacon, who had this Symbol: Let the Ternary purified by conversion of Elements be made a Monad. Lully with his three Roses, Lyon and Dragon said this: At length water is reconciled to fire. Paracelsus had on both sides the signs of the Sun and Moon, had his left hand upon the pummel of his Sword, and said with a stately and lofty mean: Separate and bring to maturity. After all these there followed a most envied man but yet a most true Philosopher. He had the world in his heart, for the universal character shone out of his eyes, he gave out himself for a Benedictine Monk, having on one side a three headed Dragon and on the other a strong Eagle. At top of his head rested the spirit of Mercury, and in his mouth he conserved the soul of the Sun. He trod in pieces the Sulphur of Saturn of unwise Sophisters with both his feet, but the Sulphur of Mars and Venus mixed with the blood of the Dragon he held with both hands. The salt of the Sun according to its preparation was not forgotten. Now whilst I considered this Philosopher with peculiar earnestness, I heard a voice crying; O

Basilius, because thou hast humbled thyself therefore shalt thou be exalted above all. From these words I understood that it was Basilius Valentinus who hitherto is so little considered.

This honourable Philosopher there followed a great multitude, which I could not all number. Each one spoke through a peculiar spirit, but yet one more perfectly than another, whence many thousands were deceived appeared thus with their writs of complaint, but they could not be admitted until first the case between Mercury and the gnashing Jupiter were layed open.

Now when these Philosophers were come together, each partly rejoyced, for they were introduced as witnesses on both sides, and on their declaration depended the whole affair. But Basilius stood a far off, and none there was who alledged him, and called him in for witness. Whereupon the Emperous and universal Governour commanded that the six named Philosophers, each under his own name should transmit unto Basilius as an impartial Philosopher, an intelligble explication upon their own writings, that he might accordingly draw up a judgement. Basilius returned thanks for this honour, thereupon collecting all their informations in writing, Hermes formed this sentence and said: That by Sal armoniack

49

Air, by the Air the Eagle, but by him the
Essentialised Mercury must and ought to be
understood, and add thereto Vitriol is SULPHUR,
ANTIMONY is SULPHUR and MERCURY, whose sweat ought
to be called the Water of Saturn for its convenient
humidity and siccity. The red Oriental Lyon is Gold,
but Sol is the fixt Earth wherein the fixt Tartar,
namely fixt salt resteth. Venus is the Green Lyon
who with her hot fiery volatile salt spirit educeth
by the help of the Lunary little world a fiery
mercurial spirit out of the cold Dragon.

CHAPTER IX

How after this sentence was fastened on the Tree of life everyone's eyes were opened, and how the Tree of life was known by its three ends standing out of which were formed diametrically and perpendicularly; and how they who had laboured in vain complain, and purpose to revenge themselves on the Philosophers and etc.

After that this sentence, as was above mentioned, was formed by Basilius, the universal Emperour who sat upon the Rainbow, commanded to publish this sentence with a loud voice, and to fasten the same in golden letters upon the Tree of Life and ensigne of salvation. Now when this was done, both of worthy and unworthy saw the mean whereby they were washed from original sin.

For this Tree was a jungy (juicy) stumpy, leaveless but not fruitless Tree, characterised with three arms whereof one arm pointed toward the East but the other toward the west. Both these arms cover the latitude of the world. The trunk of this Tree grew up from the Sphere of the Sun, and from the globe of the greater fortune straight upright. The virtue of this Tree is in its root and etc.

After this revelation there arose an universal
uproar: For those very persons who according to the
literial meaning had laboured away their money,
desired reparation of loss, and impeached the
Philosophers in an action of fraud and covin and
etc. Layed hands on them, and would no longer be
quieted with good words, but without ceasing
complained of their great pains, of the loss of
time, and their destructive loss, crying all of
them, we are undone, and are not able so much as to
provide necessarys of life. What avails us now the
knowledge when through your deceitful words we are
all brought to the utmost misery, so that instead of
materials we are scarce able to purchase necessary
bread and etc. Whereupon the tumult growing more
horrible the longer it continued, behold there arose
a most thick fog from the earth, and incompassed the
company. Some of the imaginary Philosophers who had
mingled themselves among the company of the true
Philosophers were excluded and not cloathed with the
cloud of innocency, but stood naked and bare to the
derision of all the world. These were fain to pay
the whole reckoning. Amongst the rest there was
reverend and very famous, but a haughty subtile and
crafty man, and Anti-Elias Artista. He was the least
in station amongst all the false Philosophers, but
above all much famed in respect of the multiplicity
of his counterfeit writings, which he had collected

together out of divers writers and books, and given out for his own invention. This man by the special will of God was separated from the true Philosophers by the clouds and had not his time been shortened he might have seduced even the Elect from the way of truth. For he could so bend and bow the tracts of the old Testament to his imagined philosophic work, that the vulgar and avaritious labor-ants were forced to yield him their assent. This man lived about the year when all the Gods and Governers of the upper and lower world, did signify unto us all the advent of the future last and golden age in the 9th. house of religion. He and many others (whereas they would have been retrograde in case their soul had not departed from this body) are dead before the eyes of all the world. And still to their own confusion the counterfeit Chymists must in these last times appear with the true Philosophers, and their present body be above all measure tormented by the desperate and impoverished multitude. But there came a tempest out of the Sea which carried the desperate into all lands, and there remained only a few of them standing with the counterfeit Philosophers. The same would fain have impeached the couseners upon an action of fraud and coven but they excused themselves in that they were mistaken in the understanding of the true Philosophers, promising restitution of damage. After this manner through good hope they were directed to patience, for the universal Emperor who sat upon the Tribunal had already granted them a safe conduct.

CHAPTER IX

How Mercury after he had transmuted himself into the Mercury of Philosophers.

After these things thus carried, Jupiter being first condemned, and Mercury having triumphed and the sentence being published, and it being decided in what manner hence forward the philosophic writings were to be understood, Mercury grew so insolent that he endeavoured to proclaim himself the Mercury of Philosophers, he caused his hair and beard to be clean shaved, annoyated his head with Vinegar, and the vinegar was an ornament of his head. He drew in his feet so that they could not be seen, for he mourned as a Peacock when he looks upon his feet; whereat he was astonished and grew hard like the Minera of true Saturn, for he had a Solary Sulphur and martial spirit indowed with a double nature. Yea the power of the most high is in him, and the Christian sign attesteth the universality of his Monarchy. He needs the help of no stranger, he only loves the common and true vitriolick Salt, wherein he dissolveth and becomes a clear Fountain, yea an exceeding costly royal bath, even without the help of the Sal armoniack.

After these things thus passed I heard from far a voice that cried, Wo, Wo to thee Mercury! How wilt thou now do thou most haughty spirit? Thou must down and thy feet must ascend above thee throughout thy body and life. Thou must loose thy lunary shine (splendour). If that be not done thou art disbanded by the Goddess Venus. For Behold all the benefits that thou heretofore shewedst unto Cereberus, thou hast for the most part taken up of Venus, and particularly of Neptune. O thou most miserable Mercury, what hast thou gained by the sentences, whereby the business is now come to such a pinch that sometimes there is no need of thee. But if thou shouldest, as I have already seen, cast thy skin, both heaven and earth will in the end be full of the first matter. For in all the Planets there preceded a transmutation of forms, and all this was done partly by conjunction of other Planets, as by precipitation, or sublimation of the exceeding poetious oppressed ensign in the day of the exaltation of the cross of the Lord, which the courteous Reader may with greatest attention preconceive, read, and consider farther of in the following transmutations. For my intention is that thou mayst rightly learn to know the dew Matter. The preparation is also herein described so that the serious reader hath no need to seek the same in other books. But the true understanding of my book must be mastered by an often reading. And when thou comprehendest one place, observe the same manner of explication in all leaves, and it shall never repent thee of thy pains, for which thou art to render thanks to God alone, from whom the Elucidation of all mysteries proceedeth.

CHAPTER XI

How Mars out of wrath broke in pieces his Sphere,
and at length out of vehemency of passion was
transmuted into the Shape of his King at whose side
he stood, and what afterwards was done with him.

Mars and Venus remain yet with the King, but
Jealousy creeps forth: For Mars saw that even from
the nethermost, yea from the most exceedingly
unfortunate-making common Saturn there was become so
very mighty a Monarch. This perplexed the most
valiant Prince of War before whom the whole world
quakes and trembles, from whom the Monarch himself
hath borrowed his beauty and strength, that others
were preferred before him and yet he must be subject
to the weak. Whereupon he was mad and raged above
all measure, threw the covering or Helment of his
head under his feet, laid his Sphere (spear?) upon
his terrible head, carried him so as if he would
there break it in pieces. But the noble Venus took
hold of the middle and came in the nick, whereupon
the furious Mars was astonished and became a stony
royal Mineral. For vehement passion altered his body
by way of imitation, because of the virtue which the
similitude of the thing had so transmuted the thing,
about which his vehement imagination was moved. So
powerful was the symbol between Mars and the King

that he was on a sudden transformed into his shape. And like as Diana placed horns upon Actaeon, even so was Venus inclined to set up upon Mars her own foundation and the Mysteries of her signature.

Now there is an uproar amongst the Planets, Mars had an Equil rule in respect of his hot blood, but now he is become like the most high. He had torn down the most precious Jewel from the Imperial crown, and adorned himself therewith instead of his shield. What will the feathered Mercury be able to do against the seventh Martial Monarchy, whereas this Monarchy hath in truth received all the virtues figure power and property of the old Saturn. He need never more live by the favour of others, as in Basilius, where he saith: Although the steely captain with his Spear gives Mercury work enough to do, yet can he not wholly over-power him if the old Saturn come not to his help. In sum Mars is all in all. He hath alienated from the King his best and most conspicuous Jewel out of his Crown, and put it into the hands of his well deserving champions, who constantly bear the same to his honour in their Arms, and carry it before him for his pomp.

The Jewel of this Monarch set in his crown is a most eminent Carbuncle. The other Jewel that he committed to his mighty Heros is a Crystal, and yet a Sulphur with which Luria bordered her nightgown. And behold, as soon as Mars found himself in such a form, power and dignity he was presently about to erect a new world, whereupon he awakened enmity amongst the Elements, he bid the fire to dry up the Sea, and commanded the earth to open her throat to swallow all down. As soon as the Fiat was pronounced there arose a terrible thunder and lightning, and there ascended toward heaven a very thick and most white smoke, and in this smoke was the Spirit of the Lord, who sat upon the throne, hidden. When this spirit went out as no man could stop it, but it flew and returned not again, it avoided the fire and went to its predestinated place, and it was a confused dark being, and the good thing was very little, and so far spread amongst the moist and unfit, that it could nothing profit. Wherefore the most high Emperour bid that the day should separate itself from the moist, that the soul and earth with the forementioned heavy spirit of the transformed Mars might become by solution in the fire a clear transparent liquor, and this in the same fire a fixt dry consistency: In this manner by a most peculiar providence of the most high, out of the transmuted Mars and his command a new world was made, which is serviceable to the rich and also to the poor in all respects. But alas, None is so couragious as to assail Ceberus who hath his habitation in this

world, because the same had broken in pieces the
Spears of many Knights, adventurers, so that many
Heros were slaughtered, and many also were forced to
depart with shame. Wherefore the world remained
desert and untilled, whereat the Emperour was wrath,
and he roused up to the Dragon an acceptable Enemy
as followeth.

CHAPTER XII

How the Dragon who had his habitation in this new Martial world, by the ordination of God was seized by a powerful Eagle, taken, and imprisoned; and lastly how the preparation of the true work was demanded of him.

After that the Emperour saw that it was necessary, he commanded a most powerful flying Eagle to make war upon the Martial reformed world. The same stretched his white wings wide out, and with his feathers covered the whole world, stared steadfastly with his fiery Venereal eyes incessantly upon the earth at the habitation of the poisonous hellish Dragon. He was sent out to rescue the world, and to revenge and requite the innocent blood which was shed from the claws of the Dragon. Whereupon this white Eagle with his talons seized upon the terrible Dragon, and carried him up to the high mountains of Armenia, fortified himself, and him there in a round transparent glass tower which had on one side a vault of crystal whither the Dragon was licensed and commanded to enter.

But he would not obey that command, but remained stiff-necked, lying in the bottom of the glass. Wherefore Vulcan was called to torment him so long with fire, till by a little exhalation he submitted to betake himself into the neighbouring chamber. As soon as ever Vulcan began to transgress his duty behold there arose out of the Eagle and also out of the Dragon a most horrible and very black vapour, very heavy in appearance, yea it was formed like a puralent matter mixed with some little blood, which matter had its origin from the vapour, and grew stiff before the straight entry of the neighbouring chamber. For Vulcan by the command of his Principal had not touched the same but had left it at free liberty that the Dragon might so much the willinger betake himself thither, as was also done.

After this there was very secure dwelling in the world, for the Dragon did no farther injury either to Animals, Vegetables or Minerals, but much rather promoted each according to the nature of its kingdom and custom. This proceeds hence for that the same so long staying in the Dens of the world and Earth was fed in like manner even as Animals and Vegetables. But his food is the green Earth and that most necessary spice which Neptune communicateth to us.

Now Vulcan having finished what was commanded him as to this point. The Emperour commanded that the Dragon should be delivered from his imprisonment. This the Dragon hearing looked about him toward the

Kings voice, but since he could not see the King, he asked Vulcan for the Lady that alike before stood beside him, requesting him once more to let him have a sight of her; which Vulcan having granted, the Dragon fell desperately in love with this noble Queen, so that by the impulse of fiery love he suddenly seized her and united himself with her. And as soon as this was done he be-took himself to his glass tower into his secure quarters. For (as I thought) he was not minded to quit the rape of love but he played the lover still more in the presence of Mars and Vulcan, yea in the sight of all the Planets. And this trade he drove so long until the natural Venus had intirely lost all her beauty and comeliness, and he had satisfied himself with her hot blood. Afterwards he dismissed her after he robbed her of all her royal ornaments. The natural un-transmuted Venus had by reason of Sulphur a most exceeding familiarity with the Dragon. On the otherside the Dragon had the nether part in highest estimation, and Venus the upper part of the brain of Cerberus bound under her girdle. In sum, the foundation (pavement) of Venus and the frontispiece or Cornish of the Dragon consist in the figure Analysis. She stands with her head upwards, but he downwards.

CHAPTER XIII

How the true natural Venus out of desperation precipitated herself into a deep water, and how the same was rescued by the Eagle and changed into the highest character, and what further ensued.

After these encounters happening to Venus as is above mentioned, the wan and comfortless Venus went into an empty desert place where there was a still-standing deep water wherein the mournful Venus beheld herself a long time and after that she perceived the change of her shape in the water, she was strangely incited by an inverted representation of the figure to conform herself to her shaddow by precipitation. But the flying Eagle which formerly had acquired his principal virtue from Venus perceived well that the Dragon had grieved Venus, whereupon he reached down and caught Venus by both her feet, and holding her in such a manner flew to the residence of Vulcan. But when Vulcan saw her so ill-favouredly with head hanging towards the Earth born in the Talons of the Eagle, he imagined that his Venus was dead, and knew not that the Eagle did therefore hold her head down that the water she had swallowed might run from her heart. And therefore he would neither know nor accept her, which when the Eagle saw he was monstrous angry and flew with Venus straight up to heaven towards the Sun and during the flight she by the natural warmth of the Sun was quickened, and is henceforth the subject of all wonders.

At which the Eagle rejoiced, requesting of the most high that this inverted Venus for an everlasting remembrance might be seen on the uppermost part of heaven in the face of all the world, in such form as he had presented her to her husband Vulcan, that all the world might know that that which is beneath in Venus was become like that which was highest in the true Saturn which I have here named the Monarch or Emperour. And this the Eagle requested upon this account, that men seeing the exalted Venus, how our Matter, might at the same time deride the folly of Vulcan, but he the Eagle might deserve of all an offering of thanksgiving.

This request of the Eagle was accepted, for the Emperour said:

Be it so, and as soon as the word Fiat was pronounced, Venus was seen sitting above all the Planets, and had taken in the uppermost station of the old Saturn, and also received his figure and shape. But his figure was not as the figure of him whom every man acknowledgeth for Saturn, but it was a most perfect character, that is fast shut in on every side, and their ensigne. In this sign thou shalt overcome, was at the top of it. This became the universal character excellently well. This the Heroic Mars with his knights who were present saw very well, he approached and sued as a lover in the best manner for the exceeding precious royal Jewel. But she who by the peculiar appointment of God had adorned herself with that most costly Jewel at this time knew not Mars at all, notwithstanding that he expressly put her in mind of that friendly deed which heretofore had passed between Venus and himself in courtship. But Venus who was revived from death into life (in case she may still be so called) nothing regarded all this. Whereupon Mars with due reverence was constrained to take his leave.

But Venus began, as was fit, to adorn herself with the universal sign of Salvation, and so went to the looking-glass and having beheld herself therein she was affrighted at the representation of her figure; for she no more found her ancient shape, neither knew what was done to her, but at length imagined that the mirrour must needs be so made that the same should present the nethermost above and the uppermost below, whereupon she felt with her hands upon the looking-glass, and found that the same was plain even and not hollow ground. Whereupon she

rejoiced inwardly and smiled with her shining white
countenance and ruddy cheeks, for she saw a true
change of her whole essence. Hereupon she called to
mind a wonderous fountain, which during the time of
exile she by chance once came unto, in which she
then saw herself exactly in the very same shape
whereunto she was now changed, and therefore
conceited there must needs be some prophetic spirit
therein concealed which foretold the future through
the present inverted figure.

Now whilst she was thus contemplating this matter,
she was in some measure afflicted, and said: Truly I
very well observed this my condition in the
fountain, and the vision is verified. But forasmuch
as the figure was not a meer shadow, but water also,
I still fear said the most high transformed Queen,
some greater wonder. Whilst she was thus speaking I
heard a voice out of the Air, which said: Thou
before the creation of the world wast water
therefore shalt thou also loose this thy form, and
by the help of the Eagle who rescued thee out of the
fountain, thou shalt again become a clear water.
This water called living Mercury.

CHAPTER XIV

How Mars by the help of his Pegasus endeavoured to ravish Venus and how Phoebus hindered the same.

Mars from his high rock, beheld the noble exceeding beautious new-born Queen, and having still fresh in his memory a touch of the old courtship, was well pleased with his own thoughts, and yet he received no comfort, for it seemed to him that the noble most mighty Queen was many thousand times more amiable and comely than before when he debaucht her. Thus began he to prophesy and said: Behold the wanton Venus is still a pure Virgin who has of herself the seed of her own impregnation. This and the like he said, took heart, mounted on Pegasus, and purposed ere she were aware to attack and ravish the noble Queen. But envious fortune was against him, for as he was thus hastening through the Air his shaddow was before him upon the earth. Whereupon the noble Queen looked up and perceived it was a Martial stratagem, and therefore be-took herself into a deep dark cave. And although Mars would gladly have made haste after her, yet in respect his body was harnessed he could by no means compass it. Wherefore half desperate he hastened into the next wood there to spy out his advantage. But the Queen which by this change had drawn to herself the kingdom from the monarch, was at this time, afraid of Mars, and therefore continued in the Cave. Phoebus had from above very well observed these passages, and while Mars was hovering in the Air, he shone and darted his beams on the back of Mars and his Pegasus, so that thence a shadow must needs pass before unto the earth. In this manner Phoebus discovered, and that very much to his own advantage the purposes Of Mars, for he himself was inwardly inflamed with ardent love. Now whereas this his beloved, perhaps by a special providence continued longer then pleased him in her love, he be-took himself on high, and by his golden beams descended into her deep dark cave and illuminated the same, and embraced this lovely virgin his true Mother, and fulfilled his desire. Whereupon she was loyal to him, and they espoused each other, and became two in one body.

CHAPTER XV

How Mars revenges himself upon Phoebus, stirs up Vulcan against him. How Vulcan burns up Phoebus together with the Queen in the fire, and how both their Souls appeared to Mars, and how Mars proceeds with these souls to make the same corporeal.

Mars was fain to behold all this with patient eyes from far and could not hinder it, but yet considered of a mean and way to revenge himself, and therefore with his flying horse made haste to Vulcan, Venus' husband, and brought him tidings that his wife was yet alive, and that she was a thousand-fold handsomer then before, and how she had layn a long time in dallyance with Phoebus in a deep Cave. Vulcan was ashamed and said I myself am the occasion of this mischief because I would not accept my wife heretofore when the most blessed Eagle presented her unto me. Hereupon he grew outrageous and said, Dear Cousin, lead me to the Cave. Which Mars willingly did, and therefore took him up before on his horse, and brought him to the Entrance of the Cave. Vulcan laid his ears to the mouth of the Cave, listened and yet could not learn whither the two lovers, namely the first and the last matter, were still there or no, and therefore was desirous since his heart out of strong jealousy still leaped so within his body that he could not hear, that Mars would once listen whither they were both yet present or not. Mars well observing them in respect of their inward affection so deeply in love together that they, even as if intranced, were livingly dead, gave Vulcan to understand this amity. Then prepared Vulcan an artificial firework which was made of an unkindled fire, or a fiery Air, and of a Vegetable Salt. These he mixed very well, and therewith set the whole cave on fire. But these two lovers were not aware of it, and when they understood it, they regarded it still the less, but stayed together until at length Vulcan from above threw down his prepared powder in great quantity by the Entrance, and with a most strong heat reduced them both to dust and ashes, which he brought forth and put it in his drink, which was common well-water, which got thereby a very strange taste so that Mars could not drink of it but was fain to mix it with wine. And as soon as he had poured white Wine to it, behold the clear water was changed, and out of it there was a thick and most beautiful red Essence in the hand of the inamoured Mars who reserved these ashes as reliques, and gathered them up for an everlasting remembrance, as a sacred thing. But as soon as this had so happened in the hands of Mars, he cryed out with a very loud Voice: O Venus, my Venus, thy beauty is fit for none but me alone. O fortunate chance that hath transmitted unto me so precious a treasure and etc.

And when he had spoken this with great joy, he intreated Vulcan not to take it ill at his hands, but to bestow him the blood of Venus, since Venus had in his hand (Mars) separated herself from the water, and thereby gave to understand that she will have no communication with this common water, yea and for this time ought to be rescued from the danger of water.

Vulcan (at other times none of the wisest) seeing this wonderful adventure, namely that fire, or the highly red soul of Venus had shewed herself unto Mars out of the Water, easily consented. Whereupon Mars took the remaining ashes and with common water elixiviated the hidden living soul, filtered the water, and freed the soul from the Water by pouring on Vinegar, wherewith he so long plied the soul till she shewed herself visibly, insomuch that he could very conveniently rescue her.

CHAPTER XVI

How Mars by the advice of a certain Philosopher, with the soul, of his transformed Venus, whereto the soul of the Sun was joined, attempted to awaken a new Venus, and wherein he failed. And also how by violence he wrested from Vulcan the dead body wherein the clarified body was hidden.

Now after that Mars had mastered the fiery dry soul, as is above related, he asked a very learned Philosopher how he should manage this treasure: Who gave this true resolution. Thy former Venus in her ancient form and shape shalt thou never more see, for she is now twice changed, first into the matter of the Stone, but afterwards by Vulcan with Phoebus she was adapted for the extraction of the three Principles, but thou in due time mayst see her in a shape an hundred times more beautiful. Wherefore take of her soul one part; of her spirit as much, but forget not the ashes. Put it in a close Chamber, give in the beginning gentle heat, and do in all things as thou knowest so without fail shalt thou come to thy intent. Mars did as was commanded him, but as yet he could not wholly understand all this that the Philosophers commanded him whereby to mix the blood with its proper spirit. At last of all he considered how that both of them were born out of one matter, out of one spirit and out of one vitriolic Earth. Yea Mars considered also how that heretofore he himself was changed into the first matter when Venus in his desperation held him by the arms, and thereupon the figure of a certain stony twinkling Saturnal Mineral was communicated to him as is to be read in the foregoing chapters.

Now when Mars had obtained this illumination, he considered the ashes which he left lying behind, he hastened back to fetch them. But Vulcan was before hand with him and had taken them up and carried them away with him, for it repented him that he had intrusted Mars with this most precious treasure, especially when he called to mind what he had heard and seen of many Philosophers, all whose labours he had been at. And although all his lifetime he was but a dull Ideot, yet he still retained thus much that without ashes nothing is to be performed in this high work. He had also read in an ancient book that the Phoenix out of ashes does reproduce her young ones. For these reasons he elixiviated these ashes, evaporated the water and shut up the salt of glory in a peculiar vessel, gave it a gentle fire, imagining again to hatch and awaken his Venus. He went neither day nor night from his glass, forgot to eat and drink. Mean time comes Mars into the Laboratory, found Vulcan thus busy standing before

the furnace, to whom he presently declared his
desire, told him how he had united the soul with the
true spirit, and that there wanted nothing more of
his matters but only the Salt that was hidden in the
ashes, requested that Vulcan would let him have the
Salt also, especially since there was no coagulation
nor constancy to be expected without the salt. He
demonstrated to Vulcan that the salt without the
other parts could nothing profit him, for a body
without soul and spirit is only a dead earth. By
these words the old Laborant Vulcan became wiser
then before, desired to have his wives soul and
spirit again, and said, The Soul belongs to its own
body; Now I have the body give thou me the soul
together with its seat which lived in my wife. Mars
would by no means do it but said, the spirit wherein
the soul at present resteth, which otherwise
transmits its true body to the soul and permanently
fixeth it, the same is my proper spirit. For
remember, Gentle Vulcan, said he, that here not my
Martial spirit but my spirit that was extracted out
of my changed body, as is before mentioned, can only
promote this business. Wherefore the most part is
out of my body, and etc. And whilst he was thus
speaking, he violently took the glass out of the
house, and with his Pegasus flew to this residence,
and afterwards conjoined his reserved soul through
its spirit with the salt he had stolen, and placed

the same very well closed in digestion; where when it had stood a competent time, Mars was desirous once to look into it, where he saw that all in his glass corrupted, and was as black as pitch. This terrified Mars above all measure, so that thereupon he fell into a swoon. Meanwhile by chance came in Vulcan, who incouraged him and said, the corruption must preceed; when the dark night is over then follows the White and golden light.

CHAPTER XVII

How Mars adjured Vulcan, that he would help to promote the business, and how Vulcan instructed him concerning the strange colours, which made Mars suspicious.

Mars believed all this, for he found in his mind that it could be no otherwise, and said; Dear Cousin, I well understand that thou art better experienced in the business than I, forasmuch as thou only hast ripened and accomplished the most precious Medicine of the Philosophers. I intreat therefore that thou wilt take my business in hand, for I hereby promise thee that I will henceforth live like a Brother with thee. Yea, which is more, in case the Queen of salvation shall elect thee instead of Phoebus, I promise thee to afford thee all brotherly assistance against him, upon condition that I may place on the top of my Helmet as a well-deserved Crown, that most costly Jewel which came into the world with her, out of which she was born that shines clearer than the Sun, and from which the sun hath acquired his lustre and colour and etc.

Vulcan as he gave to understand in words, was content to do all this, but purposed himself to retain the body since all must pass through his

hands. For these reasons he employed his well experienced duty, governed the fire like a Master. Mars
because of his fervent desire would not stir one
step out of the Laboratory, but stared with his
fiery eyes upon the glasen chamber wherein his
beloved was afresh to rise again from the dead. Mean
time he espied a cloud which descended down from
above, from which appeared a little small light upon
the earth, whereupon the horrible dark night began
leisurely to depart and vanish. Yea it might be
perceived that day was about to break. And when Mars
saw all this he observed further that Juno,
Jupiter's wife, but knew not for what cause, was
crept into this cloud, whom he presently knew by her
Peacocks Tail, and was afraid of some huge deceit,
especially since Vulcan lay before the chamber, and
made as though he were asleep, which the suspicious
Mars well noted. For as soon as ever he turned,
Vulcan looked to his fire, whereupon Mars drew his
Scimitar and held it naked in his hands, closed his
helmet, grasped his buckler, and remained without
turning like a statue. Mean time Vulcan was very
still but could no longer dissemble but pretended as
if he had been awakened out of a deep sleep, and
demanded, what means this good cousin, that thou
stoodest thus prepared for combar before me whilst I
was fain asleep? Then raged Mars and said with a
jealous mind, As I perceive thou art a most false

old Araitor, for behold whilst I by chance once turned about, thou hast opened the chamber and let in Juno with her Peacock, and she has hung the room with Tapistries of all colours, and adorned it with the Rainbow, and whilst I was beholding this wonder, her husband Jupiter as soon as Juno was departed came into the royal Parlour to afflect me and to ravish away the most beautiful person of this world, and as a Judge to send her back again to Phoebus.

Vulcan took great pains and could hardly appease Mars because of his settled suspicion, until he had discovered unto him the whole intrigue. Wherefore he said, hearken to me gentle brother, and give credit to my words. Juno is the Apholstress of this most noble Queen, predictress that declares that the Queen will suddenly enter and possess her kingdom, wherefore she is come as you saw to furnish and accommodate with all sorts of paintings and tapestries (as is customable in all royal courts) the royal chamber. And that thou mayst comprehend it, said he, mark carefully the example that I shall relate unto thee. In winter all things upon earth are unshapely, yea even as it were frozen to death; but when the spring approacheth the earth is again revived with all manner of Vegetables, green foliage, and invested with flowers of various colours, and when this sign is over—passed then follows the wished fruit according to the nature of each thing, and the property of its seed. Even so it is with the Philosophic Work: For if rottenness that is black death were not perceived in the Work, there would follow no germanation, Autumn or Harvest. The darkness which thou hast seen is the corruption whereof the Philosophers have written so much. But Juno with her Peacock by her greenness first declared the spring, and then by the colours the germination. Wherefore thou hast no reason to be angry, but rather highly to rejoice.

Mars held his peace a long time, at last demanded what then Jupiter had there to do in the royal chamber. Whereunto Vulcan replied and said that Jupiter descended thither on purpose to present the Scepter and Crown unto the Queen of Salvation, and afterwards to sue for the office of Marshall. For out of this matter of which this Empress is born, were all the metals likewise born. Whereupon it is also necessary that the colours of all the metals appear in the glass and that one Planet should translate and drive away the other from their office and virtue until the most permanent of them all through the beauty of Venus and the sanguinary colour of Mars shall possess the kingdom in peace.

This pleased Mars exceedingly above all measure, so that he would needs violently press into the royal hall to present the Queen with all his power, strength and authority. But the well exercised Labourant Vulcan would by no means consent, saying:

Take heed gentle Brother that you touch not the chamber otherwise we shall both be robbed of all hopes. For remember that if the budds be plucked from the Tree, that then there can be no fruit therefrom.

Whilst he was appeasing Mars with this information he opened another gate, and heated the royal hail according to his wisdom, and behold before their eyes there was a great change, and especially in the royal chamber. For the Goddess Diana was come over the glassen Sea and secretly ascended into the royal throne. She had arraryed herself in cloth of Silver; and her Ladies stood behind, not daring in regard of the brightness of Diana to behold her countenance. They held and sustained with their hands the royal Throne. But the Throne was of pure gold, and the Queen who sat thereon had golden feet, and a silver white appearance, so that the aureity which was therein hidden could not well be discerned.

CHAPTER XVIII

How Vulcan opened the last gate, and how Mars was slain by the King, such and likewise what strange wonderful colours appeared.

Mars had observed all this very well, but dust not ask Vulcan anything further, for he saw him busy in opening the last gate. When it was opened the fiery flames or heat assalted the Kings Palace in all Quarters, and there appeared many thousand mirrours, and each mirror represented the beauty of him that was to reign without end. Mars looking roundabout could not espy the Queen who had taken upon her the shape of Diana, but found in her stead the most wise King Soloman sitting in his glory, and therefore requested to Vulcan that he would further and permit him to enter into the transparent chamber, that so he might hinder King Soloman to take his well beloved wife. Mars had hardly spoken out what he was saying, when he so perceived by the red cloathing that his old Venus with her hot blood was still present, which now arrayed herself in cloth of gold. She continued not long in one shape, She was now before his eyes in form of a man and presently after in that of a woman. Vulcan well saw that Mars resolved even by force to press into the royal throne whereupon he led Mars by the hand, washed him from head to foot cleansed and annointed him, and brought it so to pass that in he came. And as soon as he was gotten into the chamber wherein the wonderful and strange changes had appeared, Mars was seized on so as that without the least defence he unobservedly swooned before the well exercised Vulcans eyes. Now after that the most mighty Hero Mars out of love of his Venus had lost his Life and was slain by the spirit of the Dragon with which the royal hail was 1/4 part filled, King Soloman revealed himself in his glory, sitting upon his triumphant chariot and peacefully reigning over the whole world. The red and silver coloured Sea was subject to the king and willingly yielded up its riches in his time. In summ the King was mighty in peace. For he as the last Monarch by the help of the old Dragon was become Master of the beauty of Venus, and the glorious valour of Mars.

For the Dragon was tamed, and was no more as heretofore filled with hatred, and enmity, but was above all measure gentle to the inhabitants of this world, predicting unto them salvation riches and health, and etc.

CHAPTER XIX

How Mars after he was subject to the King and also
spoiled of his purple garments, sought and found
opportunity to revenge himself upon Phoebus who
formerly hindered him from bringing his beloved
Venus to his will, and also gave occasion to this
his misfortune, servitude and captivity, and the
loss of his purple robe.

Although the Dragon prophesied peace and riches to
the whole world, and all things everywhere seemed to
be in a quiet condition, yet the same lasted not
long. For there was still a peculiar enmity between
Mars and Phoebus, whom Vulcan upon Mars' notice had
before taken in adultery with Venus and with a
peculiar thunder bold and fire-work had wounded to
death and consumed so that Mars in whose presence
this passed, hoped that it would only have effect
upon Phoebus and not upon Venus, confiding that she
would soon rise again safe and sound as out of a
deep sleep, and alone possess the government of the
world. But he was in his imagination hugely
mistaken. For as soon as he was introduced by Vulcan
into the royal Hall, wherein Venus had given up
herself unto the King, with purpose to rescue the
Queen, Behold he was there presently laid hold of by
the transformed Venus (whom I here name the King)

and he became subject unto her. Mars when he saw it could not otherwise be, kept close to the King, who was singularly well pleased with his fiery eyes, and hos purple coloured under-garment which he stript him of and adorned himself therewith. All this Mars very little valued but only thought how he might revenge himself upon Phoebus, and after upon the cheating laborant Vulcan, who brought him in. Wherefore he spoke to the King, and said; Most mighty and invincible King, it is true and known to us all, that thou art the heart, life, and glory of the world, thou art replenished with the spirit of the double nature, thou shinest like a Carbuncle in the night and darkness which from the beginning covered the world. But Phoebus, said he, He inlightens the day itself, so that the day by the interposition of a gross intransparent body in the place where Phoebus' lustre cannot arrive is to be esteemed as darkness, and in this manner his lustre may plainly be distinguished from the day light. Wherefore said he, I counsel your majesty to bethink of a mean and way how you may get into your hands the Mantle of Aurora. For my part I am confident I could come by it without much trouble, and in case your Majesty give me commission to this purpose, deliver it into your hands in a short time. This proposal pleased the King mightily well, wherefore he commanded Mars to do as he had said, and etc.

Mars well knew that Phoebus would not suddenly be able to forget the lovely relish of the fruit which grew in Venus' garden and which he had newly tasted. For he very well knew the hot nature of Phoebus, but specially then when Phoebus in shape of Apollo imagined to overtake the exceeding fair Daphnis and missed, he had very plainly espied his Nature.

Now the crafty and revengeful Mars had observed and out of his Watch tower wherein he was heretofore lodged had seen that the most amiable Heliotropia had often times between whiles walked alone at the bottom of the Hill, and amongst others for some hours had recreated herself in beholding her own shape, and that she made a looking-glass of a crystalline spring, and fell in love with her own beauty. So that for very love she was like to dye, for she burned with her own fire, which Cupid had enkindled, for as much as he had impressed upon the heart of the most beauteous Hellotropia the sign of love, viz, the character of Venus so bright clear and visibly, that it shone forth like a bright shining star between her alabaster breasts in her own shape and true figure. This same star and figure Heliotropia had not before this rightly observed: but now when she stood before the fountain she espied the character. For the water which she used as a mirror, represented the same much fairer and more glorious then it was in itself. Wherefore the inamoured Heliotropia desirous to lay hold of and kiss the star, her feet suddenly slipt, and behold the fountain closed over her head, and Heliotropia lost herself therein. This Mars had well noted and resolved to serve himself of it against his enemy, as hereafter followeth.

CHAPTER XX

How Mars with cunning, and the relation of a wonderful adventure concerning Heliotropla, wherein the Chymical Mystery is expressly and visibly delineated, revengeth himself upon Phoebus, and how Phoebus sinketh in the fountain, and becomes a ferment of the last Monarch.

Early one morning as Phoebus came over the Sea, drawn by his unweired Horses, and had directed his golden rays toward the Martial hill, to draw up the various coloured Vapour from the crystalline fountain wherein the spirit and life of the metamorphosed Heliotropia was hidden with which he was wont to adorn the Rainbow with all manner of colours. Mars approached unto him and with his hoarse voice said: O thou most illustrious Phoebus, if thou knowest the adventure that hath happened in these parts, thou wouldst not only be amazed at it, but with all thy might wouldst revenge and punish the injustice which has been exercised upon the criminal himself. For whilst the self-enamoured Heliotropia as was her custom, was walking in this valley by the appointment of Cupid had artificially prepared out of certain salt matters generated in this Mountain, and at length sat down and fell asleep; suddenly had the lewid Cupid by the instinct of Venus so wounded and surprised her with love toward her own shadow, that she was forced more and more without ceasing to behold herself in this fountain, which she so long did till at length a most heavy sleep fell upon her. In this deep sleep Cupid over-took-the most beautiful Heliotropia and resolved to purchase her most beloved treasure while she yet slept. Having therefore done his desire he gave her a chain of Emeralds whereat hung the portrait or figure of his Mother Venus most artificially wrought. Now this image was the occasion of the unfortunate chance of Heliotropia. For when she awakened, she presently according to her ancient custom turned her face toward the fountain, and when she perceived the strange amiable shape of the effigies between her alabaster breasts, she stooping toward the fountain fell in and melted away like salt in water. And after a little while I saw Cupid at this fountain, and forasmuch as I could perceive, 'twas he that contrived this adventure, and enchanted Heliotropia in the Fountain, that henceforth he only and alone might enjoy her beauty at his pleasure. For in the place where heliotropia last sat, as soon as she was drowned there presently sprang up a wonderful flower. By this mark Cupid understood where he might find his beloved.

At these words the colour of Phoebus' countenance was changed, and his whole body was inflamed like fire, and he stood in such a transport that he was not able to speak. Wherefore Mars spake further and said: Be comforted dear Phoebus and give ear to what I shall advise thee. I well see that thou art inflamed and inwardly and outwardly inkindled; wherefore if thou whilst thou art thus in thy fiery form wouldst plunge thy self into the fountain, the fountain must needs dry up, and the fair Heliotropia could no longer remain concealed from thee, but must then needs become thine. For if thou beest so powerful as by thy rays from far to dry up great waters, what shall not thy presence be able to perform?

Phoebus could no longer refrain but believing all that Mars had related, on a sudden without any deliberation plunged himself into the fountain, which had the nature like royal water in which all red colouring stars are wont to melt away. Wherefore Phoebus also was overpowered by the force of this water, so that he was forced to become water, but without his prejudice. The most vindictive (revengeful) Mars knew this right well, who had long before experienced the effect upon his Venus, and therefore to deceive Phoebus he contrived and found out this invention, in consideration that heretofore

Phoebus overtook his transmuted Venus in a Rock and as is above mentioned, had his will of her at his pleasure.

Here the gentle Reader and lover of Art ought to take special notice how wonderfully Mars, after that out of love to his imagined Venus (I say imagined for she was metamorphosed) by the subtilty of old Vulcan with whose wife he heretofore played the wanton and is still much inamoured of her, was punished and reduced into servitude; did first revenge himself on Phoebus by hindering him from the rape of Venus when he would have ravished her and carryed the booty away himself. And it is also here further to be observed how the Monarch of the Philosophers became through the tincture of Venus and Mars a most general, universal and afterwards through the Mantle of Aurora, and the presence of the corporal solar Phoebus was fermented, so that the otherwise most spiritual fixed Monarch did last attain an Earthly quality and affinity of the baser metals. This ferment is the mean between the metal to be tinged and the tincture that is to tinge. After this manner I have treated through all the Planets on purpose so to reveal the next matter, for it pleased not otherwise my Creator. But in all places there is to be found a several Metamorphosis of the stars, how the base, mean, worthy and most worthy Planets by several manuals, fortunes and misfortunes, and yet by the appointment of God were changed into the matter that is next to us. This nearness must be reckoned from the Planet, as if it went backward. And by that means in this the first. After you come to a running Mercury, and lastly to a thick vaporous fume, which at all times breathes (exhales) in the mines and caverns of the Earth.

There is no more difference between the lord and the servant, between the nethermost and the highest, between darkness and light. For I have here in this world reduced all the nethermost Planets into one and the same thing, so that the nature and property of them all may be truly signified with and through one only character. The Planets of which I now discourse, are the noble and ignoble metals which are inclosed in our Elemental world, as it were in their matrix: I say in our World.

Now the first matter of which I have so much spoken and treated, has also her seat and habitation herein. Through this matter, if the same be freed from its earthly moles or habitation, the Philosophers have made their most blessed stone. The matter of which we here treat, is a metallick form, but the reason why it is accounted a mineral is because the due matter wants the form. The matter

prepared desires this form no less than a woman does
a husband, both these are in their metallick water
solved, coagulated and fixed, as I have seen. But
the thing wherein the metallick water is concealed,
has a symbol of the greater and less world, and is
only to be known by its simple signature, and the
fore-described transmutation. But for the signatures
and characters, (although I have already
sufficiently declared them in my treatise of the
Magical Elements, and the annexed chymical Alphabet)
how they are to be known in and by all the Planets
before your eyes, and in a fit and due time to be
brought to light I shall hereafter (if God permit)
discourse, and delineate the matter after another
fashion, and by the Philosophick compass bring it
into so fine and convenient a form through that one
only word complica, so that those who have any
faculty of considering cannot but, yea must against
their own will, understand it. Even as I in some
places have been impelled by the Philosophic spirit
against my will to write too clearly. God grant thy
the salvation of thy souls be herein always minded.
AMEN.

Now I will proceed to the Moon who receiveth no
Metamorphosis.

CHAPTER XXI

How Diana was by Cupid deceived and gotten with Child, and how she was forsaken by her Nymphs because she had lost her virgin form or crown and etc. And how by the conjunction of the uppermost Moon with the nethermost Moon which Diana had on her head she was changed into the Saturn of Philosophers and etc.

I even now made a failure in that whereas out of the baser metals, as our Saturn, Jupiter, Mars, Venus; I brought forth unto the mis-shapen common Mercury such very strange and unusual things, but there made no mention at all of the Moon. Therefore it is fit that I now introduce them to the most chast Diana, which for the most part is delineated naked and uncloathed. Let the gentle Reader always observe this: that the Planets by a peculiar contrariety love or hated have always diminished, exalted and bettered their condition. For after this manner they became the best of all metals, viz, the first matter whose virtue and mean unregarded figure or form, I value higher than the best gold upon earth. For in this thing is hidden the seminality of gold, of which I dare here speak no farther. For behold the night approacheth and the Moon comes from the East with her horns downwards and therewith threatens my left eye, (for the Moon governs the left eye even as the sun governs the right eye). Now to preserve the same I must be-take myself out of the wood into the plain field, and behold how the Moon will appear unto me, and in what figure she may and ought to be delineated.

Now when I had gotten into the plain field, I heard behind me the moving of a water, and I observed that it was the chast Diana who after her virgin custom was come down to Bathe herself in this fountain. I looked wistly upon her with singular pleasure, and perceived that all her Nymphs by a peculiar adventure were fain into a very deep sleep. At last I understood the occasion of this sudden sleeping for I heard a good way of the voice of an inamoured youth. I hid myself partly out of fear, partly out of desire to understand the issue of the matter. Mean time whilst the chast Queen also out of curiosity hearkened to the exceeding delicate voice with a singular relish and attention, her eyes thereby became dim and the longer the cloudier, so that she with her Nymphs was overtaken with a profound sleep. And when I turned my eyes toward Diana, by the reeds I espied a beautiful white swan which made this so lovely melody. He had hanging at his neck the Quiver and Darts of Cupid the God of Love and conservation of mankind and etc. From whence I conjectured that it must needs be Cupid himself, who plays his tricks with all flesh but could never gain or compass anything of Diana.

The God Cupid by the instigation of his Mother Venus (for as much as he burned without ceasing for the love of Diana) had changed himself into a swan. For after the musick was ended I saw the inamorate in his true shape and being. His mother Venus had let herself down upon the earth in a troubled dark cloud, stood upon the shore of the water, called her son unto her and bid him drink out of both her breasts.

This she called Virgins Milk: For by metamorphosis she was become a virgin. Afterwards she again reached him a very heavy draught which was mingled with her virgin blood, and with that also which was hers, as a mother while she appeared in her old unchanged shape. As soon as Cupid had drank this off, he was in a strange taking, and after a little space he acquired all sorts of colours, burning in the water like a fire, during which both the clouds and Venus vanished. Cupid bent his bow and addressed his arrow so as Diana perceived it not: Whereupon the leud Wagg repeated his shot again, which allighted too near Diana, who started as if she had been ready to wake. But Cupid was sorely afraid that

if he should again bend his bow and let fly, that
Diana might thereby receive some harm and be too
much inflamed and etc. He therefore hastened to the
thick reeds, and as soon as he got into them he
began again to sing after another manner. Whereupon
Diana awaked together with her Nymphs, but the
Nymphs being awake knew not Diana who stood before
their eyes, for the hot Cupid in shooting had beaten
of and shattering in pieces her Virgin Lunary crown,
and Diana too by the affright and smarting of the
fiery dart of Cupid had gotten a red and discoloured
face, whereupon she was not known by her Nymphs,
which afterwards went up and down the woods to seek
their Goddess. Diana thought and well saw that some
trick was put upon her, but knew not what was done
to her until at length she beheld her shadow in the
water whereupon she wept and lamented uncessantly,
cursing the fountain in which this happened to her
and others beside had before been beguiled. Now
although Diana stood on the water in a very
afflicted condition, and being robbed of all joy,
was thinking of means and ways for her rescue, yet
it was her highest consolation that unknown of her
own she might yet a sufficient time continue in the
same place. Meanwhile she observed the most
beauteous Swan, and thought to herself, surely this
must be some God? Who spoiled me of my virginity,
and gotten me with child. Whereupon she made as if

she were sleepy, and lying down just in the same place where she lay before, she famed as if she were fain asleep. The Swan observing not what Diana had in her heart, approached to her as heretofore, at which Diana took fast hold of the Swan, but he covered her eyes with his wings not willing to let his face be seen. But Diana holding him so fast that he could not escape, the Swan changed himself into his true shape, and said: My beloved Queen, I am not come to afflict or disgrace thee, but to honour and glorify thee so that thou shalt bear solary children. For thy predestination is such that the stately hot sun shall be derided, humbled and exiled by the cold earthly Moon. Thou and thy children shall shine clearer then the Sun himself when he stood in his highest lustre. The Sun vouchsafed that foolish request of Phaeton, and while Phaeton rode about with him they came too near the world and burnt up the world, and all that dwelt therein, and I also am a child, of the world, and now by the help of my transformed Virgin-Mother, the cause of the most high wonder and its multiplication. And behold said he further, the world is at present without comeliness and outward ornament, but thou and thy children, which shall be born out of my hot blood which I also have received of my Mother, shall have an heavenly lustre and an earthly glittering, and also a particular heavy spirit, shall fill the heaven and be cloathed with the beauty of the earth. To that end shall thy head hence forward be adorned with an imperial crown, and the world shall be delivered into thy hands. Now when thou and thy children shall be born upon the Solary triumphal chariot, and ridest about the whole world, and by thy moderation fittest and promotest it to all kinds of births, thou shalt be put in mind by the character of the world, which to that end thou hast in thy hands, to span the world. The Sun which thou carriest on thy chariot shall shine through thee, and thou shall impart thy bright fire to the sun himself. Now when he had said this, he gave. his Bride a fresh kiss, whereupon she forgot herself, letting loose her hands which by her fast grasping were already strained, and Cupid fled upwards on high, leaving to Diana for a pledge his fiery enamoured heart.

Now Diana surely knew of whom she was gotten with child, but still was in doubt of Cupids promise: For she well knew that lovers do some times promise such things as are impossible for them to perform. With these and the like contrary thoughts she tormented herself, and then cast her eyes upward, and then again turned them to the fountain, and when she had espied her appearance therein, she beheld that the Moon that formerly stood on her head, did again stand in its place, but it was all guilded over, and the Moon which shone in heaven shone down from above into the water, and both the Moons each with its horns made a pleasant well-proportioned conjunction in form of a round fiery globe, out of which grew the Tree which bore the fruit of Eternal life; And I heard a voice from heaven which said unto me: This Tree is called the Tree of Salvation, the Tree of recovery of Eternal Life: The spiritual water which was brought out of it is called the spirit and water of life, a living fountain and etc. Wherefore take heed that thou neither comest too near nor violatest either this Tree or yet the Earth. Now when I well understood this, I turned mine eyes to the circle out of which this Tree in my sight so suddenly grew up; and I beheld that it was incompassed with a golden crown wherein all sorts of costly solary stones were set, amongst which the Martial Ruby, and the venereal Emerald had the pre-eminence; for by their Virtue Diana was to bear solary children, and to accomplish her predestination as is hereafter to be read.

CHAPTER XXII

How Diana after that she was as is above mentioned turned into the first matter being gotten with child by Cupid, brought into the world a Salary birth.

When Diana observed the time of her delivery grew nigh she gave a loud shriek, and there went a fiery lightning out of her mouth, and she lost herself so that I could see nothing more but the fore-mentioned fiery globe, whereon grew the ensign of Salvation, which instead of a rail was incompassed with a golden Crown, until at length the heaven became very bright and clear through the lustre which Cupid brought with him out of the Cabinet and warehouse of his Mother. Herewith he adorned his beloved Goddess, and carried her on the Chariot of Phaeton, where she brought forth a new Solary offspring, that was so like the Mother that one could not well be known from the other. Unto this new birth were given the Sun in the right hand and the Moon in the left, for he had power to burn up the world, to quench and afterward to make it fruitful. For these reasons he had the cold moist Moon and the warm dry Sun as two opposite qualities, in both his hands, thereby to signify that gold and silver do proceed from him. This birth was in form, bigness and figure, weight and lightness comparable to the whole world, for never was there found a greater and more glorious offspring upon earth.

To this regent of heavenly light and earthly salvation the glory of Mercury joined itself, bringing with it exceeding large wings, which by the venereal property and the hardened lye of the briny Ocean was exalted into a very beautiful white colour, sparkling clearer then the light itself, which shined in the left hand of the new born saviour. This so exceedingly precious offering Mercury by command of Jupiter and all the Planets brought, requesting help and assistance in the name of them all, for they were hardly handled and injured by their earthly enemies and were subjected to corruption. But the cause of their mortality was their own contrary quality and the discord of the three principles, whereupon they could never be secure in their state unless the most high, who with fire and cold changeth, conserveth and filleth to all kinds of births, the world and all that therein is according to his will, would take their part, and erect a true harmony in them, and by this means they might withstand their enemies. With these wings which mercury had brought as an offering, the redeemer (assister) of the Planets was very well appeased. He could with the same notwithstanding his exceeding great and mighty weight, like a spirit through Vulcan mount up on high. To his ascending there was perceived in his mouth a wonderful Balsam, which proceeded from him through the pores in form of an oil. And this by the wise and understanding in Physick was with most profound reverence gathered up as an holy thing. Of this blessed, coagulated, essentialized, purified and incered oil, Mercury as soon as he had layed down his commission brought with him for every Planet as much as was needful. And when they by direction of the ordained receipt which was fastened upon every little glass, had used the said oil, they were sensible of their own melioration in seasonable time, but they imagined they should presently without any precedent ablution of original sin and imperfection which cleaves to them, as it were all in the innermost depth of their centre, arise and appear in their former complete state of innocency and perfection. Wherefore it was declared unto them, and I know not from whom, thus spoken: The time is not yet come that through the most high in a moment the weak and imperfect can pass into its highest perfection, but the spirit only endeavoureth, in the virtue of its magnetical nature to mix itself with the corporeal Planets and by the power of their own souls to carry them gradually till at length they arrive to the highest degree and equality. In which equality all the Planets finish their course namely in the point of the heart of the Lyon, in which very place nature hath erected a golden Column, whereon stand these words written in the Arabick, Hebrew and many other Languages: Go no further, here nature rests in the mineral Kingdom, here let the Traveller stop, and the Artist make haste and etc.

While I was beholding this Pillar with singular attention, Mercury stept in and writ these words upon fine Virgin Parchment, explaining the same to all the Planets, and at last comforting them altogether. He said: Ye must all for this time pass into your fortune and perfection through great sorrow, anguish and trouble. But if the assister with the strength and reddiness of blood thirsty Mars and the fair Venus shall appear in his beauty, which by the exercised Vulcan is so ordered that he is able to withstand the hellish Pluto with these furies and lay him for dead, and also tame the infernal Cerberus, and take him to thy inheritance, then shall ye through him in an instant be freed from your mortal weakness, which can no sooner be done unless the Sun expiate his guilt, who by the unexercised unlearned Phaeton's government set the world on fire, whereby the inhabitants and giants fell at enmity and strife with the Gods, so that the mighty of this world stormed heaven and laboured to pluck down the Gods from their throne. Whereupon as we heard above the Planets dispatched Mercury unto the new born Saviour for help and assistance, presented unto him the most sumptuously compacted wings, to move him by so costly a gift to compassion and etc. as it also happened. For as reduced Mercury brought back for each so much of the rich oil, that everyone was considerably strengthened thereby.

CHAPTER XXIII

How Phoebus fell down backward from the Chariot, How Pluto assalted him with his hellish poison, and how he was turned into a three-headed worm. Also how his brethren forsook him and delivered him to Vulcan and etc.

Now after that Diana by the contrivance of Cupid and furtherance of the Martial Venus, was ascended to so high dignity, I heard these words from the four quarters of the world declared to all reasonable creatures, as me thought it was the voice of the Goddess Diana the bringer forth of secular salvation. She said thus: I had perished except I had perished, there is no evil through which the most high creator of all things cannot effect some good. For behold I was weak vessel, a tender wavering-minded Maid, loved chastity, punished all those who attempted upon the honour of me and mine. To Acteon because he Espies me naked in the bath, I gave of my Lunary horns, and he was unknown by his most faithfull hounds, torn in pieces. Wherefore I have in anotherwise contrary to my will been deceived. For whereas I was most highly in love with chastity and punished immodesty in others, I fell by my own curiosity whilst I had all my faculties possest with the pleasant song of the divine Swan,

which was Cupid himself: and after that chance my
Nymphs also knew me not, and I feared least they
also should serve me as Acteon was served with his
beloved dogs. But mine innocency and proper
predestination exalted me to this place, where no
mischief can hence forward touch me.

After I had heard all this there arose a very
terrible voice out of the midst of heaven, and the
voice was the voice of one that judgeth and
pronounceth sentence, the signification whereof was
this. God resisteth the proud, hardly shall the
righteous be saved, where shall the wicked appear.
Hereupon the Sun was darkened and fell backward into
a deep fountain, which wanteth no fire and Pluto
fell upon him bringing with him the venom of the
hellish Cerberus; he had in his left hand a leathern
bag filled with all sorts of deadly stinking powders
upon which was written Dragon and Stone-serpent
powder. Mix them and let them flow and etc. With
the forementioned powder Phoebus in exile was
nourished and the food went to the Soul of him, his
living spirit departed from him and his former
beautiful form was no more present. He was so
changed by the forementioned meat and fiery drink
that his own brethren knew him not, stood a far off
and beheld him as a despised worm lying on the
earth. And the worm which they saw, had before and

on him the end, and also in the middle an head.
These heads according which was to outward
appearance were not like one another, for the fore-
most had its original of Sulphur, the hindermost of
Vitriol and the middlemost of Mercury. This
middlemost head had eyes before and behind and
looked with singular attention upon the foremost and
hindermost head, and out of its ears went two hands
which laboured to bring the hindermost and first
head together. But there came an angel named Uriel
who emptied a fiery Vial upon the middle head so
that it was fain to quit the fore-most and
hindermost head and fly into the Air. For there grew
wings out of the place where the hands stood, which
would not for-sake the Sulphur and the Salt. These
wings bore the middle part into its own mercurial
kingdom, from whence it shall again return over -the
glassen Sea to enter upon its office, and then will
the most high send forth his spirit and unite both
the forsaken solary parts viz. Sulphur and Salt
which were changed into a worm, with itself in the
most subtile parts. The Ten Lepers shall obtain of
his hands the divine Nector, and all thereby be
cured and his brethren who know him not shall fall
down and worship him. His brethren heard this voice
ascending out of a deep fountain, and therefore said
Wo Wo to us if our Brother be lord over us, he will
remember all since we forsook him in his greatest

necessity, and afforded him no Brotherly assistance when Diana by the help of Cupid, chased him out of his kingdom and etc. his hand will be always hard against us. Therefore it should be my counsel said the blood thirsty Mars to rid our Brother Phoebus out of the world and so we are secure of him. But Jupiter said let us not soil our hands on him, it is better that we deliver him to his eldest Brother the double-armed Saturn who has power to free him and also to retian him. Whereupon he was delivered to Saturn who by the instigation of Vulcan cruly tormented him, would fain have stript him of his purple mantle, but it would not succeed, wherefore the ravening Saturn was enraged, returned again to Phoebus, and said thou shalt and must whither thou wilt or no accomplish my will, it is fit that first of all without any excuse thou pay me thy turn-key-fee, and hangmans wages, but whither thou dyest or livest is alike to me, I must in both cases be the better for thee. Whilst Saturn was thus harrassing the wretched Sun the most high regarded him with gracious eyes, and the Sentence was so lenified that Phoebus should not die, he be banished the land, and because he burnt up the world, for an Eternal memorial receive a brand-mark. Wherefore Saturn brought the brand-iron to the place wherein execution was done, and impressed it hot as it was upon the patient Sun, commanding him to avoid the

bounds of the kingdom, and always to cloathe himself
with a grey coat. For he knew now no more anything
of high-mindedness was reduced as it were to nothing
by humility; seeing he was made of nothing, he had
taken on himself the form of the first matter,
seeing whereas otherwise he was the last matter of
all the Planets. In sum the last is become the first
like as the first became the last, for they both
proceed from one, and also both hasten to and aim at
one, wherefore one may be turned into the other.

CHAPTER XXIV

How Phoebus came into AEgypt and was chosen king, and how his brethren as was prophesied, humbled themselves before him, and how Phoebus changes his name and nameth himself the revived Cross-bearer.

After Phoebus had duely expiated his sins, envious fortune was forced to forbear his tricks, and he who likely might otherwise have easily despaired constantly thought on the good promise touching his person, and thus with a good courage went continually forward towards Sunset, and came into a wilderness wherein dwelled many learned pious Monks who always imployed their mind in the search of the mystery of nature, and thus by the knowledge of the temporal, so much the better to understand and more inflamedly to love the Eternal indestructible things. These expected the coming of the holy Ghost who might impart the knowledge to them. And whilst they were once gathered into one, they saw Phoebus in his present form coming afar of who bare up the Cross upright upon the forehead of his face. When they saw him they thought to themselves, this must needs be a holy man, and lover of God, who is so martyred for the name of Christ, and has imposed upon himself openly to bear about the mark of his Redeemer all the days of his life. Wherefore they all made haste unto Phoebus, and after they had well interrogated him, they understood that there must needs be some great mystery comprehended in this adventure, intreating Phoebus to stay with them during the time of his life, and fundamentally to inform them of his adventures. This he promised readily and did as they requested and instructed and taught them the mysteries of nature, and he faithfully revealed unto them whatsoever was in the closet of nature which hath its being in the centre of the world. But they were all fain to make a vow that they would reveal these communicated mysteries to no man.

Here fortune changed herself and Phoebus changed his name, naming himself the revived Cross-bearer. None whilst the revived Cross-bearer conversed in this world and quartered amongst the wise, they became from day to day more prudent and mighty, so that the Princes and all the kings of the earth stood in fear of them. And that they might all have one head and lord, they elected the revived Cross-bearer their King by inheritance and guarded him so well that no stranger without their fore knowledge and will might see him: For they feared least so mighty a Monarch might be craft be surprised and taken from them, or otherwise be injured.

During the time that the revived Cross-bearer was
King in AEgypt there arose a terrible earth-quake,
so that the whole world was astonished thereat; yea
all mines and caves of the earth fell together, and
the Catarrhacts of the world being stopped took
another issue, from whence arose great necessity and
affliction amongst the Mountain-Gods, and a
miserable lamentation was heard amongst the Planets.
For the court of Saturn was filled with an inexhaust-
ible multitude of Waters: The Palace of Mars and
Venus was so layed waste that the place where they
were, could not be found. The Caves and Grottos
together with the Bath wherein Diana sojourned, were
faln together. And the spring soaked up Mercury
toward the South, and Jupiter toward the west had
suffered the most dammage, Well knowing that they
deserved such punishment in regard of their Brother
Phoebus, they all came to confession and said
unanimously, had we then afforded help and
assistance to our Brother the Sun when Cupid by
Diana, otherwise unfortunate, chased him from his
throne, he might now have had power and money to
build up the Palaces, and to cause our shattered
Members to be cured. During the lamentation Mercury
stood upright on high on his feet, and said: Dear
brethren, there is a most mighty King chosen in
AEgypt whose wealth and understanding famed all the
world over. The same is above all measure liberal

and neighbourly, wherefore I counsel you to dispatch me with my Brother Saturn thither. For I know when he shall see the old man halting thus pittifully, he will then demand, who is this honest reverend man? To whom I will then answer: Most invincible, most understanding and most gracious king, this old man is my true brother, there are yet two brethren and one sister beside, who incompassed with sadness stayed behind, which cannot yet bear thy countenance, and etc. All that Mercury proposed to his brethren in the aforesaid manner pleased them altogether very well; wherefore mercury with Saturn must needs undertake the journey. Now when they came into AEgypt the wise men of the land would not permit them to go unto the King, until at length the King out of his chamber espied the shape of Mercury, and caused enquiry to be made what they both had to do at Court, and when the King understood that they were both come a great way to speak with him, there was therefore presently audience vouchsafed to both the envoys. Mercury had his Scepter adorned with the Serpents in his right hand, and stept foremost in and making his due reverence with bended knees, said: Most mighty and most intelligent King, we are five Brethren and two sisters, all born of a most powerful royal Mother which heretofore had the whole world in her power whence we also bear in the arms a character of the world, and etc. But the Mother so

ordained yet the most beautiful of us should possess
the glory and pride of the kingdom with this
proviso, that Diana at her pleasure might chose a
royal seat for herself in the four corners of the
Kingdom, the rest of us have obtained our
inheritance in the valleys and caves of the hills
under the earth.

Now it is known unto thee, O most bountiful King,
that through the universal Earth-quake much dammage
has been done in all the world. Amongst others the
misfortune has light very hard upon us five, so that
all my Brethren together with the loss of all their
goods as they were in their caves were pittifully
wounded in body. Whereupon we thy servants are fain
to beseech thee that thou according to thy laudable
custom dismiss us not without favour.

The King commanded Mercury with his old Brother
Saturn to sit down, asking the old man his name, and
whither he were born lame, and requiring Mercury
circumstantially to relate the adventure and
condition of all his brethren. Whereupon Mercury
answered; My Brother Saturn came into the world
defective; is hated by us all, for if he be
throughly heated and come too near any of us he
destroys him and etc. He eats us and our children
out of unsatiable appetite: But we honour and

respect him; For Medea transmitted unto us a prophesy, that by Saturn we should all become Kings. Whereupon we dare not punish him in his wickedness.

At this the King laughed and looked wistly upon Saturn, imagining the old man would know him, and call to mind how he formerly tormented him, when he was called Phoebus, by the contrivance of Cupid was beaten from the Salary triumphal Chariot, chased out of his own kingdom and delivered up to this Saturn as is more largely mentioned above. But the old Saturn not daring, an evil conscience urging him, to behold the King, the King commanded him to declare his own harms. Whereupon Saturn retreated back and said: I am but a dead man, this most mighty King is my own Brother, I was formerly his Executioner, I know him by the sign of the Cross, and the brandmark. Now it is fulfilled that we should all worship him.

Whilst he thus cryed, Mercury knew his Brother as it were by the ruddiness. For although outwardly he had put on another shape yet inwardly in the center by reason of blood he was still the same. For the same that he was before his fall can be again out of him produced and etc. Mercury fell down upon his face beseeching grace for himself and his brethren and Sisters. Whereupon the King commanded him to fetch to court his other five brethren which stayed behind in a running Coach. Mean time Saturn must stay behind as a pledge.

As soon as Mercury was returned to his brethren he related to them altogether the whole adventure, but they trembled and quaked, riot daring to appear before the King; and yet they were more fearful if they stayed away, that the King driven by wrath might cause Saturn to be slain, and at once deprive them altogether of their glorious hope which they founded upon the prophesy of Medea. They chose rather to dye then live without hope, and therefore got up, and in a short time came before the King.

Now when the King had well viewed them together he demanded where their sister Diana, and the Sun stayed. Then answered the just Jupiter, and said (for he and those who had stayed behind, did not altogether know the King as Mercury and Saturn did). Most gracious King, Phoebus in regard of a certain crime and misfortune which proceeded from him, by the providence of the most high Creator of Heaven and Earth was condemned to the fire. But Vulcan, though assisted by Saturn was not able sufficiently to quell him without Pluto's Recepe. They indeed plagued and Martyred Phoebus, that at last they grew weary themselves; whereupon the sentence was mitigated, and the same who was condemned to death, was only thrust out of his kingdom and brand-marked, and as we understand he passed over Sea toward the Sunset, and by a storm the ship was beaten to pieces against a Rock, and the wild Savages took him and burnt him alive in that fire which proceeds out of the Mountain Aetna. Diana the Earthly Moon our dear sister was in her sleep gotten with child by Cupid, and carried up to heaven, and with her kindred is become immortal, and that taken upon her the place of the Sun. Now we had imagined by this Saturn thy Brother to inherit the Earthly kingdom, but it will not succeed with us; thy fortune subsists in the highest misfortune. Saturn is the greater in fortune, through him we must pass into the emendation and happiness.

The King was pleased with the discourse and wisdom of Jupiter above them all, and said: The wickedness which ye executed upon your Brother Phoebus sprung from the haughtiness of you all. For you endeavoured to hinder and frustrate the providence of God, and the promises made to your Brother, and yet by your own sinister action the predestination of your Brother was promoted. For behold all of ye. I your Brother Phoebus am at present through my own backward fall and by divine providence to all your profits turned into the true first matter and saturn

119

of the Philosophers, whom hitherto ye have not known. What think ye now, do ye imagine that your wickedness may remain unpunished. No by no means. I have sworn that I would drink the blood of my Brother and Sister. For my sister Venus was most guilty of, and the occasion of Diana's first getting with child, and that she was after qualified to bring forth in my Chariot, a salary Heir, who drove me and thrust me out of my Kingdom whilst I was yet Phoebus: But at present I am a wonder above all wonders. And as he was about to explain himself he was changed before their eyes, as is to be read in the following chapter.

CHAPTER XV

How the revived Cross-bearer shewed himself in his former shape when Pluto with the Hellish Venom turned him into a three-headed worm, and how Vulcan delivered him from his curse and original sin, and annointed him with the blessed oil which Mercury brought with him for the health of his Brethren, and inflamed him through the blood of Mars and Venus, so that he was reduced into a true universal most general Medicine.

As soon as the revived Cross-bearer was resolved to satisfy Justice he grew wrath in himself, and behold he stood in another shape namely in the shape of the worm with three-heads, which as is above mentioned comprehended in themselves the three principles, and by the manuals of Vulcan were revealed.

For as soon as he the revived Cross-bearer, took upon him this shape, the well-experienced Vulcan appeared, and by his knowledge first revealed the Salary Sulphur, secondly its Vitriolick Salt, as also its own mercury; but the spot or curse of sin be separated, as a most rank poison, from the three principles; and when the first and the last, namely the Sulphur and Salt were cleansed, he dissolved them both, each in a several vessel in their proper

Mercury, and when they were become a liquor he
united them both, and they mixed as water with
another water, and became a mercurial water. But
afterwards the Sulphur became one with the body and
began to take upon it an equal regiment, and to make
the spirit bodify with a convenient fire and at
length most permanently to fix into the best
medicine, at which many strange wonderful incredible
beautiful colours appeared.

Now after that all the colours were passed, and a
brownish red mass did unchangeably rest alone in the
bottom of the glass, then the revived Crass-bearer
appeared in his banished shape, he opened the median
vein of Mars and Venus, this blood he drank of,
whereby his wrath was mitigated.

Vulcan rejoiced at this, seized on the revived
Cross-bearer purified his fiery three-fold Soul,
mixing the same with the blessed oil of the
Philosophers and gave it to the three-headed beast,
which now again was become one thing, to drink
thereof so much as he had need of, and when the
draught was consumed he repeated the same drink to
the seventh time, and when this drink was dried up,
the three-headed Monster was delivered out of his
captivity, and the corporal purified Sun greedily
received it, for there was none amongst the Planets

more deserving it than the Sun. Now when he had lain sometime in the lodging with the Sun, it was discovered all the world over that the most mighty assister of all the Planets had entered upon his government; for he layed his fingers upon the wanton Venus, and behold presently her heart was moved with Mary Magdalene, and she was on a sudden changed by the power of the most high redeemer, and sparkleth like the Sun.

The most blood thirsty Mars was sensible of his brothers grace, and whereas he (Mars) before this always strove against the fire, and with fiery sparks assaulted Vulcan who hammered him, he gave himself so humble and gentle under the hammer like the melted common Saturn, he was inwardly and outwardly arrayed through and through in cloth of Gold. Jupiter, Saturn, and Mercury all eat of the golden food with their Brother set before them, and thereupon forgot all the pains they had undergone during their former life, they became by the aforementioned food immortal as long as they forbear unripe grapes and the true Magic Elements and whereof may be further read in its place in the treatise of the Magical Elements.

CHAPTER XXVI

How Saturn being about to try whether his brethren were in truth amended and turned into the Sun, was by the command of the Emperour annointed, and in respect of the exuberant medicine tinged his Brother Jupiter into the Sun and etc.

When the Saturn of the vulgar saw all this, before his eyes, his cold frozen heart burned with new envy like fire, so he went out and chid his brethren for their folly because they were so easily credulous, saying it was a sophistication, and praying Vulcan for experiment sake to send one of his Brothers to be examined who had given out that he had experienced transmutation in his own proper body. Whilst he was requesting this with impatiency, I heard a voice out of the point of the crown of the Emperous saying: Be it so, Be it so. Whereupon Vulcan stept forward took hold of Saturn and placed him upon the Hellish Throne of Pluto in an inundation of fire, and when he was enkindled Vulcan would have also brought them thither of whom Saturn was in doubt in respect of the transmutation, to try whether in truth they could stand out in the examination. But he who already had shown grace to Mars and Venus gave Vulcan a drop of his blood, with command that he should privately laud and annoyst

Saturn who expected his brother in the midst of the fire. This Vulcan accomplished as he was commanded, and as soon as he had brought it near Saturn, the Balsam touched and moved the heart of old Saturn, and he was changed not only into the best gold but he had also retained with him from the, abounding medicine a tinging property and power, by virtue whereof even the rest of the imperfect Metals can attain and be reduced to perfection as hereafter may be read.

Mean whole Mars appeared in his Martial habit saying to Saturn, know thy self, let everyone prove himself and then judge thou of the truth of Transmutation. Saturn well observed that Mars mocked him, and then contrived how he might prove the children of Mars, Venus, Jupiter, and Mercury, and the Moon, whether they also inherited anything of the royal blood of their Ancestors or not. For these reasons he said to his brethren and sisters. I see and acknowledge by your form there must needs be another kind of understanding concealed in the Philosophic writings, that I perhaps am not the Saturn of the Philosophers. For my part as I in my halting form was wandering up and dawn and put many in great hopes, I have never as yet been profitable to any and etc. But be that as it will I would fain know, said he, whether the benefit which the Assister, the revived Cross-bearer has conferred on you does stick close to you and yours, or not. The petition of Saturn, Venus first of all vouchsafed, but Jupiter was before hand with her and went into the fiery flood to the old Saturn. At his entrance they both wethered (breathed) and Jupiter through the abounding medicine which Vulcan had privily applied to saturn, was touched, that he with his examination grew stiff in the fire. Vulcan spying this laughed heartily in the presence of all the Planets, saying, As I perceive Medea was not altogether out of her story when she gave so many and great hopes to the assembled stars in regard of Saturn. For he has tinged even Jupiter himself into the best gold. This virtue the most high caused Vulcan to apply to him in the flux when he was to examine his brethren, by which he was changed into a medicine, for he had taken to him more Of the incombustible oil then was needful for his health, and this chanced well for Jupiter wherefore Jupiter returned thanks to his brother Saturn, who besides his intention and by chance had done him such a favour, which was not from the power of the common Saturn, but from the farce of the Tincture itself, which benefit other metals with more profit may enjoy by the revived Cross-bearer.

CHAPTER XXVII

How the two worlds strive one against another, and out of them as out Of two contrary natures the spirit of Mercury is made. And how the Lunary world is provoked to the fight by Neptune and otherwise, and so qualified to the contention, and etc.

After that such changes as the courteous Reader hath understood had happened, I by chance be-took myself into a very still Quarter to consider the adventure and the signification thereof, and while I was encompassed with my deepest meditation, beheld at that time there was a great and exceeding beautiful picture let down from heaven, in which all which had happened in the manner related concerning the Metamorphosis of the Planets, was compendiously and artificially delineated, and a King with an imperial crown held the painting with both his hands. This King had his power in his voice, for he spake out of the point of his crown, and earnestly commanded me to copy the painting which he shewed me, and in its time to fasten it to a Column which he would then shew me. He hardly made an end of speaking this when it became so dark that I was terrified thereat, and imagined no other but that the last day was come. Now being encompassed with very great anguish, whilst I was looking about how I might get out of this darkness, I saw on my right side a wonderful and terrible Enginc (cnsign?) in heaven that had a shape like the world. Now as I was looking toward my left side I observed likewise another ensign which was in all things like the former, only one seemed nimbler then the other. But these worlds were by a peculiar wind driven one against the other as if there were an enmity to be brought to light between them. There was a vast horrible and deep abyss between them, and one abyss called the other to fight for both their spirits, which yet were of one original, and by the will of God were to fight one with another till the appointed time. The contention arose by reason of preferency and right of primogeniture. Wherefore both the worlds in virtue of their predestination clashed hardly one against another, but lost themselves both in regard of the darkness. Whereupon the world on the right hand called the brightness and strength of its Son, namely of the Sun, and commanded its son that he as to his natural Mother, should yield his light to serve it, and continue first and unmoved in heaven, and afford it all necessary assistance, whereby its adversary the left world, might not perchance by the help of the Queen of night lye hid, or escape unhurt. As the right world had commanded, so was it also done. For the Sun let himself down by a golden chain upon the world his dear Mother. But the Mother mightily rejoyced in the obedience of her Son, embraced her Son, at which the abyss opened itself and swallowed up the brightness of the Sun with all his glorious appearance. Whereupon the left world rejoyced who had called her daughter the Moon to her assistance, who also to help her Mother slid down by a silver chain, and stood by her Mother. The Moon had seen how the Salary Mother had carried away her dear only Son into the utmost perdition and into the Abyss of the center, whereupon she endeavoured to hide her self under a cloud. But the womb of nature she spake out of the Abyss, Fear not Diana thou Daughter of the left world, for thou hast found favour with God. Whilst the spirit was thus speaking, the Abyss of the lunar world opened itself by descent into a fountain full of water, and when I had delivered it out of the cold water I perceived that the spirit or vapour was turned by water into a weighty moveable water. This water was obscure and not transparent like the ice which is found in the Salary world, there where like Sun is at the hottest, but only glittered in regard of its high preparation and like precedent ablution like a most clear looking glass. On this mirrour beautifully green illustrious Venus was inamoured. Neptune observed the secret courtship of the lunary glittering world, wherefore he sent unto the green Venus a transparent crystal, which by the

evaporation of superfluous moisture he had so formed
and reduced into a glassy-Sea, whose signature was
as the signature of the most necessary spire. It
retained with it the last from the Palace of
Neptune. This Neptunial crystal and most green
Venereal Emerald were delivered over unto Pluto with
an ancient process in writing, how to use it. The
title of the Process was: Up with the Volatile twice
or thrice. Out of the mirrour like clear Fountain
shall there be a mist, and of the mist shall there
again be born the snow in the higher region.

CHAPTER XXVIII

How the Lunary glittering light is exalted and clarified, and how the same became a Hydra and an Eagle, and also how thereby the world gave up its spirit which appeared in the shape of a very heavy mist or cloud.

Pluto had this Process long under his hands, could not manage it, called Vulcan, asked advice of him, who said: That which God by such a wonderful adventure hath joined together, that let no man separate but rather promote it, wherefore they shall be one; The above mentioned Smaragdine Crystal and the glittering Light which by the help of Vulcan was born out of the Abyss, ought to be locked up in a royal Transparent Hall, where they must so long continue, till it be learned what thy Gods are resolved by such and adventure to Effect.

After that it was thus concluded (shut up) the three above mentioned spheres said. Let there be made a powder, Let there be made a fire, Let there be made an Exhalation, And when all was still Vulcan made fire under the Chamber and behold a white beast with two horns exalted itself from the Earth towards heaven, whose wings were two clouds, One proceeded out of the Sea, the other out of a sower mineral juice and served the horned beast through their innermost ability. The clouds were the food of this beast of which the beast ate so long until it became a very poisonous white Hydra. And when Vulcan a fresh and for the third time fed the Hydra with the offering of Neptune and Venus, the worm then changed itself, and it became a bird which with its wings could cover the whole world. It had its Talons like an Eagle far wide asunder and bit with its beak upon the Pavement, and with its talons seized on the world from the sunrising and sunsetting. And the world together with this flying venomous beast hid itself before my eyes so that I could see no more but only a very heavy thick vapour. This moved itself by the benefit of Vulcan to climb on high but the heaven hindered this so that it was of necessity forced to pass sideways into the next cave, which had no issue but the Entrance well served it to go out at and etc.

CHAPTER XIX

How out of the above mentioned cloud the Philosophic Virgin is born, and how she distributes her milk and blood for the comfort of the needy. And also how to bring the soul into the Metallick body, namely salt, and etc.

When this cloud was passed I saw a beautiful white virgin ascending from the earth, which pressed her breasts and made out of her virgin milk a wholsome butter, wherewith she purposed to restore life to the dead. For she cryed with a shrill voice: Die ye living, arise ye dead. After I had very attentively viewed the above mentioned virgin, I saw that after her hard pressing, which she would not leave off, a thick blood flawed out of her breasts, for there was no more milk present. This blood defiled the butter, whereupon Vulcan endeavoured to take away the butter, but the Virgin cryed and said. The milk, the butter, and the blood are all good, but each of them hath its several operations, which shall be revealed to no man until he shall learn out of what I am born, and which my Father and my Mother were.

When I heard this I was affrighted and was afflicted thinking perhaps it is granted to none to learn this. But when the Virgin observed that I was sorely troubled, she said unto me with a smiling mouth: The grey Dragon is my Father, and the white bird is my Grand-mother, and both these are Sisters and Brothers children. The blood I have from my Father, the milk hath its original from my Mother, the hermaphroditical world, whereof this milk also through the virtue of the Sulphureal soul left behind, must become blood. Herewith the virgin pointed to her neck where the soul was hardened into a palpable substance. This she named, the Jew who stained his hand in the blood of the Redeemer. The Colour was like an Artificial Cinnabar, but heavier and browner in form, wherefore the preparation of this thing was unknown to me. I hearkened to the voice of this saluton Virgin, and poured the flowing Virgin upon the stiffened animated substance and the milk turned into blood, and the blood was full of life. For the nourishing metallick milk became one with the spirit of Life, which also opened his mouth and said: Two suffice to awaken a living body, but the third conserveth it.

After I had Understood this I gave unto the Metallick soul a metallick body, and I bound the soul with its body by virtue of its own spirit which as they, was likewise metallick: This union was the cause that the former Virgin from whom salvation ariseth unto us, of more sorrow and grief was fain to cloath herself from top to toe in a black Mantle, and as me thought it mightily repented her that she had so cleansed herself, and that her most noble juicy whiteness and redness should so come before the unworthy. For she said: Wo to me, wo to the world, wo to the Mother which bore me. And whilst she was thus crying she changed her beauty into the shape of a horrible Wolf, but yet retained her tender white virgin breasts. She lay as all Wolves in a deep bushy valley, from whence that most devouring beast watched for food and nourishment suitable to its nature, and etc.

CHAPTER XXX

How the Mother of Nature committed to the Virgin her Children, and how the Lion and his Sister are not to be overcome but only by the double infernal thunder.

When the beast was now satiated, there came an intelligent ancient Matron with long garments and much gravity. She had two sons, and committed to the She-wolf both her sons to bring them up till her return. And having layed the children to her breasts, the old woman turned herself into a vapour, which pierced into all the neighbouring hedges, hills and stones, from whence I observed that it was the Mother of Nature which would have her two metallick offsprings be brought up by the She-wolf as a metallick wet-nurse.

For this mother of nature was very well acquainted with the above mentioned Virgin, notwithstanding that she was changed into a She-wolf. She discoursed also with me, and said not the outward but the inward is that which the Artist ought to seek. The same further said this, which is well to be taken notice of: My Virgins are learing Wolves, Dragons and Serpents, and etc. my Son is a roaring Lyon. He hath sucked the breasts of his Sister, and is thence become so fiery and pugnacious that no beast upon earth can overcome him except the double infernal Thunder with the substance and durability of Bacchus came too near him. Thereby will this my son and daughter loose all their powers, and take their flight to me as their Mother, and etc.

Thereupon all was still, and I thought on the double infernal thunder with the substance of Bacchus, but could not search it out, and therefore cryed out with a loud voice. O Jupiter thou that in thy hands bearest thunder and lightning, shew me the mysteries out of which thunder and lightning have their original, whereby I may fright away and force Wolves, Lyons, Dragons and other monstrous beasts.

After I had thus prayed, the heaven grew stormy and Jupiter spake from far. Two contrary things are known by last and small, when two contrary spirits fight one with another then arises a great tumult in the circumjacent Air.

This I also understood not, wherefore out of discontent I cryed also unto the God Pluto. I considered that since the Thunder was infernal it must be granted to Pluto alone to revile it. Then was I heard, for he presently granted me too fiery spirits. The one was wan and meager as death, but yet above all measure long and nimble. But the other was heavy thick dry and fat, and looked on all his body like a man sick of the Jaundice. Out of this throat went a very poisonous vapour which stabs living creatures and refresheth Bacchus. The first spirit was overlayed with twice as much strength as the second. Wheresoever these spirits went in which me I was held in suspicion, and I stood with them in great fear, and therefore I would have them no longer by me, but made them a quarter under a shed without doors, that no body might take notice of them. Now in this house there was a ruinous Cellar wherein there inhabited an old grey wolf, which was fain every day to be fed with a red fox and a yellow Cock, which the Master of the family to be complained of promising me same thousands if I could help him so that either the winged wolf might die, or at least leave the Cellar. Now when I had well heard all I called my spirits to counsel, which said that there was a royal treasure hidden in this Cellar, and that the Wolf had devoured many wealthy Passengers together with their money, who in the times of war had gone down into the Cellar to rest. And for as much as gold and silver could not consume in the Stomack of this beast, the treasure was still there. These above mentioned spirits offend themselves, and said; If thou wouldst let us in we would deliver all the treasure to thee. I presently said, Go and bring me the treasure.

CHAPTER XXXI

How Bacchus and Vulcan jeered the Laborant; How nature teacheth him to offer unto Bacchus, and how the Lyon having thrice conquered, at length fighting with his father was overcome.

As soon as I had let out the spirits I heard behind my back a great laughter, and turning myself about I saw that Vulcan and Bacchus stood by me laughing and jeering at my folly, saying, with dammage shalt thou be wiser, thou that wert ashamed to worship us as Jupiter and Pluto, surely 'tis better to ask then to go away with shame and loss. The spirit does nothing alone, but there must be a protection on the other side which must guard the spirits. This Protector ought also to stand in good agreement with Vulcan and etc. If out of thine own head Bacchus thou canst hit this go to thy spirits and to the Cellar. This affected my mind, for I had left my spirits in the Cellar to no purpose, and they lay as if they had been dead there, and the wolf had put on a Lyons skin, and nature appeared to me in a mist, and told me how I had sinned against Bacchus and Vulcan, how I should go from this place toward the south, there she would entirely inform me How I should move Bacchus to my assistance by an offering. Whereupon I hastened to that place and betook myself to a

Forest. Now as soon as I was come into the forest I there saw a Lyon who fought by way of offence against a Dragon which is wont to be called a Lindtworm. Then thought I to myself what need has the Lyon to Fight with this Dragon, when yet all Lyons descend from the old Dragon, and also no Lyon alone hath ever been able to overpower a Dragon. This must needs be a great presumption which engaged this hereunto. Whilst I was encompassed with such thoughts the fight on both sides was so vehement that it seemed as if the Lyon would be vanquished. But I was grieved for the Lyon, and yet I durst not approach to help him until the Mother of nature out of the air commanded me to rescue the Lyon whereupon with my Scimitar I chopped off the Dragons tail, with which he fastened himself about a tree, so that thereby he was torn in pieces by the Lyon, and his blood was sucked out, from whence the Lyon was as beautiful and gentle as any man in this world. Mean time nature stood betwixt me and the Lyon saying, this Lyon must Duel thrice with the Dragon his Father and each time be victorious. But at the fourth time he must begin an invincible battle against his Father, not in the Air, nor in the water, nor yet in the earth, but in the fire. And in this fight as well the Father as the Son shall so wound themselves that the pavement shall overflow with blood round about, which she bid me to gather up with express commandment that when both of them were provoked by Vulcan to the fight, I should continually strengthen them with a burnt sacrifice of fat Vervains. It was said to me, the water of magnaminity is a dry drink for the Lyon and Dragon, by this the Soul of both of them will be inflamed so that they will not be aware, when they bleed, and when no more blood comes, also neither father nor Son is to be seen. Then take it again mix and melt, then, all is open. Separate the extremes, and take heed that you conserve the life of the third, for two have the third very subtile in them. Which third is made corporeal by the spirit of Mercury, and at last acteth upon the extremity of its former and latter, and conjoining them most firmly, and he himself is conjoined with those things which are to be conjoined.

When the understanding slow-paced Nature had done speaking this I demanded how I should appease Bacchus by an offering. Then I received this resolution: Bacchus loves strong drink, if thou canst well separate the same from its cube which dominates over the ternary and cause the three principles to be delivered to him by Vulcan in a golden vessel, then shalt thou thyself see what is grateful or unpleasant to him, the same shalt thou improve to thy advantage; For that which he takes to himself is agreeable to him, and what he lets alone take away, and etc. So shalt thou attain thine intent.

There is here more contained in these Chapters then the Courteous Reader imagines; in as much as amongst other things the true Saturn is everywhere to be found, and like a picture is exposed to all the world. I have in this Treatise written in a peculiar manner concerning the common Saturn, and all the earthly Planets and have so long bended them backwards, and forwards, crooked and bowed them that one has been changed into the other, the highest into the lowest and the lowest into the highest of all, the first into the last, and on the contrary

For in this metamorphosis of the metals, the true matter of the Philosophers is so conspicuously and clearly in all the changes of the Planets, brought to light, that any man that has but wit, whither he will or no must needs thereby know the first matter together with all necessaries, especially if the same hath taken into consideration the offspring of the world, the conjunction of certain planets, as also the wondrous fountain, its representation and reflexion. All which I as a looking-glass unto the transmutation of Metallick forms by a divine command and impulse have revealed to this unthankfull world.

I know that many Philosophers, I say many titular Philosophers, chymicall writers, and collectors of processes, out of envy will readily gainsay it, especially those who have understood nothing at all concerning the miracles which have already been done by me in chymical labours. But the cause of this miracle is not that I have perhaps purposed to make myself great, but was therefore done that I might manifest my experience and vocation, and thereby bring the erroneous into the true way. A man should enquire of him who has written out of his own experience, and receive of him who himself hath obtained grace of God. For these reasons I have in divers Countries with one part tinged some hundred parts of Saturn into pure gold. This tincture is a hindrance that I shall not be impugned by the envious, and it is also a warning before all the world, and etc. Wherefore let not the Courteous Lover of Art suffer himself to be seduced by strange false labourants, and process-venders, and particularily, but always be thinking how he may search out the foundation whereon the whole building resteth. Then shall his eyes be opened, and he shall be able to distinguish good processes from false ones.

He who from this my faithfull hearted information learneth the first matter, let him be hereby assured

as truely as I through God live, that he may most
certainly learn and find its preparation also, the
universal menstruum, the true irreducible reduction
into the first matter by the first matter, the
conjunction and fixation into a most precious
medicine, together with the augmentation and
fermentation in this little Compendium.

A MEMENTO

Courteous Reader,

I have not without considerable reasons handled here in this treatise the concentrated Ens and etc. by way of Metamorphosis. For I, in regard that all Philosophers have written obscurely, and that their style is unknown to the world, am minded by this my way of writing to open the eyes of the understanding of all the lovers of truth, whereby they may understand not only me but also all true Philosophers. He that can prefer and choose the signature of all things and not the thing itself, to him its not needful to apply nose-saddle, nor to seek my keys out of my complica, but they which are endowed with a fuller understanding I direct to my chymical Alphabet, which because in my treatise of the Medical Elements it was not altogether printed with the right characters, I will anew cause to be printed, translated into Latin, whereto I have also conjoined the figure of all the Planets, the most hidden philosophic character, the nose-saddle through that wholesome imperative complica. Then will the ignorant see how long they have trodden the first Ens under foot.

The concentrants also who with a malignant mind do endeavour falsely to cry down the true tincture, saying that the tincture is merely a concentrated thing out of gold, and can tinge no higher than the weight of the Sun weighed out of which the tincture was drawn. These dull ignorants judge as a blind man of colours, they have not all the days of their life seen a concentration of the Sun, much less made it, how then can they know such a thing, and how can they verify it. But thou, thou lover of knowledge, take heed of Satan, who whilst thou resteth and delayest to go forward according to my instruction, casts in so poisonous a deadly weed upon the sowed field whereby the seed shall be kept down and can came to no perfection.

I had purposed in this Edition to add hereto an exceeding short explication of all the chapters: But since the wickedness of the world is constantly lying before my eyes and ears, it is high time to forbear and abruptly to break off, least that after they have eaten of the tree of light, they plunge themselves into everlasting darkness. But the honest of which there are very few, which fasten this my faithfull admonition to their hearts, and yet partly stumble and halt from one author to another, on them I will set my nose saddle, and so complicate, that they thereby shall be ruled as with a bridle, and shall be forced to give assent to the truth. But these are they which hesitate not out of malace but out of defect of grace, and because their time is now coming and hard by that they by virtue of their predestination must come to the knowledge of the truth, therefore I must with my new instrument namely my nose-saddle, draw the same unto me, and force them as it is written compel them to enter. But the ungodly which would fain be esteemed before the world, and write against the truth which yet is unknown to them, cannot abide such a nose-saddle upon their nasule snouts, and understand not the word Complica at all; how will they then be able to distinguish the figured from the figure, and to choose the sign for the thing signified? These notwithstanding their wicked conversation, I desire like all Thomas's to take warning that they look well to themselves least perchance they be not beaten by the strong Atlas, who threatens them as if he would throw them with the world on his neck and therewith shatter them to pieces.

But thou most politic black fox with thy Hypocrisy and crafty inquision, stand a far off, stay till roasted pigons fly into thy mouth, and thou wilt stand to the end of the world, there where thy unbelief, because thou hast known the good and out of laziness hast neglected to effect it, shall condemn thee. But I thank the most high who hath imparted his grace to me and commanded me thus to write. Wherefore I thank him and none other, neither secular nor spiritual, and say from the bottom of my heart. Glory be to God above who governs the stars of Heaven.

To the Sun I adjoin a triumphant Standard which was shown to the most mighty Caeser in a cloud with this inscription: In this sign shalt thou overcome. This flag is a sign of and belongs to the Sun.

Catch he that Catch Can.
Catch least thou be Catcht.

Beleiv't or thou beleivs't amiss
The world at which I aim is this;
Neptune & Venus cause to fly
The Snake, which else beneath must lye.
By Niter and by Sulphur hurt'd
Thereon, Mars does enforce the world.
Bacchus does the soul enfold,
And the Sprightly vapour hold:
The spirit does rule, the soul cooperate;
Earth, namely Bacchus seat, do not forget.

THE END

A Word from the Publisher

Thank you for purchasing this small work from The R.A.M.S. Library of Alchemy. During his lifetime, Hans Nintzel was dedicated to the identification, acquisition, study, retyping and, when necessary, translation of what he considered to be the most important known works on Alchemy. Hans was assisted by his sparse network of fellow Alchemists, all members of the Restorers of Alchemical Manuscripts Society (R.A.M.S.). I was an active member of R.A.M.S.

My goal is to publish all of the works originally available through R.A.M.S. as photocopies. To facilitate this, I have chosen to have the books professionally printed. I also have a few titles that I intend to add to the original R.A.M.S. Library, selected by strict criteria established by Hans.

The works from the original R.A.M.S. Library are republished by R.A.M.S. Publishing Company in the collection, "The R.A.M.S. Library of Alchemy," with permission of the Estate of Hans W. Nintzel.

If you have a work on Alchemy that you believe should be a part of the R.A.M.S. Library, please contact me through R.A.M.S. Publishing Company.

Philip N. Wheeler

www.ingramcontent.com/pod-product-compliance
Lightning Source LLC
Chambersburg PA
CBHW080813180526

45168CB00006B/2424